Empirical Agent-Based Modelling – Challenges and Solutions

Alexander Smajgl · Olivier Barreteau
Editors

Empirical Agent-Based Modelling – Challenges and Solutions

Volume 1: The Characterisation and Parameterisation of Empirical Agent-Based Models

Springer

Editors
Alexander Smajgl
CSIRO Ecosystem Sciences
Townsville
Queensland
Australia

Olivier Barreteau
Campus Agropolis 361
Montpellier Cedex 05
French S. Territ

ISBN 978-1-4614-6133-3 ISBN 978-1-4614-6134-0 (eBook)
DOI 10.1007/978-1-4614-6134-0
Springer New York Heidelberg Dordrecht London

Library of Congress Control Number: 2013947733

Printed on acid-free paper

Springer is part of Springer Science+Business Media (www.springer.com)

Series Foreword

Scientists require methods to undertake their investigations; methods are the tools of their trade. Applying the same method in different empirical contexts allows for comparative studies and, thereby, for insights that inform and develop theory. Applying methods differently in different contexts does not allow for comparative work and contributes little to research beyond the particular case study. Nor does research contribute much if methods are applied without the necessary scientific rigor. Therefore, talking about methods means to concentrate on rigor and some level of standardisation. Most methods have been improved over many decades to improve their robustness. Underpinning assumptions are contested and some methods gain scientific rigor while other methods perish. Ultimately, scientists with blunt tools will not be able to progress knowledge.

Agent-based modelling is a relatively new methodology and able to be employed by many disciplines, similar to statistics or mathematical modelling. Largely developed by computer science this modelling methodology thrives as there are substantial demands for cross-disciplinary modelling. A particular advantage arises from its ability to specify algorithms in largely qualitative, logical structures. Such rule-based designs are similar to how social scientists describe cognitive and social processes of human decision making. This ability to formulate disaggregated human decision making processes in a simulation model provides agent-based modelling with a considerable advantage.

Given this advantage, agent-based modelling is gaining currency in empirical situations, only comparable to statistical methods transforming analytical procedures in many disciplines during the 19th century. Several universities contemplate the introduction of agent-based modelling to their coursework, in particular Economics, Sociology, Ecology, Computer Science, Engineering, and trans-disciplinary studies, such as sustainability related topics.

Empirical agent-based modelling aims to reflect a specific real-world situation and often involves stakeholders that relate to this context. This distinguishes empirical agent-based modelling from hypothetical or theoretical agent-based modelling. At this early stage, empirical agent-based modelling is mainly implemented for simulating real-world systems related to, for instance natural resource use, transport, public health, and conflict. Decision makers increasingly demand support that covers a multitude of indicators, which, in many situations, can only be approached by using agent-based modelling; in particular in situations where human behaviour is identified as a critical element.

However, empirical agent-based modelling faces significant challenges that can be grouped into a list of large segments:

1. How can behavioural dimensions be characterised and parameterised?
2. How scalable are human and social variables?
3. How can modelling processes be designed to effectively support decision making?
4. How can empirical agent-based models be validated?
5. How can social networks be implemented in empirical situations?
6. How can bio-physical environments be implemented?

This series aims to bring together some experiences and solutions for these challenges in empirical agent-based modelling. Creating a platform to exchange such experiences allows comparison of solutions and facilitates learning in the empirical agent-based modelling community. Ultimately, the community requires such exchange and learning to test approaches and, thereby, develop a robust set of techniques within the domain of empirical agent-based modelling. Based on robust and defendable methods agent-based modelling will find a broader acceptance among research agencies, decision making and decision supporting agencies, and funding agencies. Currently, many steps in empirical agent-based modelling are ad-hoc choices without a robust and defendable rationale.

This series aims to contribute to a cultural change in the community of empirical agent-based modelling. But it requires researchers to be transparent about their choices. Without the necessary transparency, methods and outcomes cannot be compared and the community foregoes the opportunity to learn about what to do and what to avoid when implementing an agent-based model in empirical situations. Unfortunately, the current culture is dominated by journal papers that fail to document many critical methodological details.

This series starts with the characterisation and parameterisation of human agents because the ability of agent-based modelling to specify behavioural responses of human agents is pivotal for its current success. Thus, assumptions made for specifying such human behaviour in the simulations seems a step of high importance, requiring tested and robust techniques.

I do not see the volumes published in this series as final compilations documenting final recommendations. Instead, these volumes should be seen as snapshots requiring updates as the community learns from comparing and testing the necessary steps of empirical agent-based modelling.

Foreword of the Editors

This volume on the characterisation and parameterisation of empirical agent-based models aims to contribute to better tested ways of how to inform assumptions on human behaviour in agent-based models. Facilitating such learning in the wider empirical agent-based modelling community requires overcoming a few challenges, of which we want to raise a few.

First, we wanted to connect and involve large parts of the empirical agent-based modelling community. While we believe that we involved a broader base than the previous framework developed by (Smajgl et al. 2011) there are still pockets we have not sufficiently engaged with. However, we hope that this book is seen only as one step in a longer process of improving the methodological robustness of empirical agent-based modelling. Second, there are many ways to cut a cake. When revising the necessary framework to step through the characterisation and parameterisation process we required a structure that is sufficiently generic. The difficulty was the consideration of iterative approaches. In some ways all examples that were discussed for this book had some iterative element. But different modellers used iterations differently, which needed to be ignored, to some extent, to allow for a generic framework. Third, we wanted to assemble examples from different pockets of the empirical agent-based modelling community as we believe that some groups have made more progress with some techniques then others. Therefore, the potential for learning seems largest when connecting these groups. Fourth, trying to step towards recommendations means to identify particular situations in which the guiding principles hold. Clearly, there is no set of rules that holds independent from the modelling situations, in particular the availability of data. We had hoped to present examples for all theoretically possible cases but for many cases we did not find an appropriate example, with authors ready to test the framework. However, we hope that the examples we collated reflect current reality in the empirical agent-based modelling community, with some cases (or situations) more often encountered than others.

This volume is made of 13 chapters. The initial chapter sets the scene with the detailed description of the framework and its various possible implementations, the methods at hand, and a tentatively exhaustive set of cases of possible modelling situations for empirical agent based modelling. Then chapters 2 through 11 provide examples for empirical agent based modelling. The final chapter discusses the efficiency and the robustness of the proposed framework as well as a first attempt to draw recommendations on selecting methods for empirical agent based modelling. This last objective is clearly still in its infancy on the basis of the small number of examples gathered here. But we hope that this book will contribute to the emergence of a community nurturing a database of empirical agent based modelling cases and provide working and explicit examples to newcomers to this approach of modelling close to their own cases.

Alex Smajgl
Olivier Barreteau

Contents

Contributors

Alfons Balmann Leibniz Institute of Agricultural Development in Central and Eastern Europe, Halle, Germany
e-mail: balmann@iamo.de

Olivier Barreteau IRSTEA, Montpellier, Montpellier, France
e-mail: olivier.barreteau@irstea.fr

David Batten The Commonwealth Scientific and Industrial Research Organisation (CSIRO), Melbourne, Australia
e-mail: David.Batten@csiro.au

E. Bohensky CSIRO Ecosystems Sciences, Townsville, Australia
e-mail: erin.bohensky@csiro.au

G. Deffuant Laboratoire d'Ingénierie des Systèmes Complexes, Irstea, Aubiere, France
e-mail: guillaume.deffuant@irstea.fr

C. Doscher Faculty of Environment, Society and Design, Lincoln University, Lincoln, New Zealand
e-mail: crile.doscher@lincoln.ac.nz

Andreas Ernst Center for Environmental Systems Research, University of Kassel, Wilhelmshöher Allee 47, 34109, Kassel, Germany
e-mail: ernst@cesr.de

Elizabeth A. Fulton The Commonwealth Scientific and Industrial Research Organisation (CSIRO), Hobart, Australia
e-mail: Beth.Fulton@csiro.au

N. Gajaseni Chulalongkorn University, Bangkok, Thailand
e-mail: gnantana@chula.ac.th

Lei Gao CSIRO Land and Water, PMB 2, Glen Osmond, Adelaide, SA 5064, Australia
e-mail: dr.leigao@gmail.com

F. Gargiulo Laboratoire d'Ingénierie des Systèmes Complexes, Irstea, Aubiere, France
e-mail: Floriana.gargiulo@gmail.com

Armando Geller Scensei, LLC and George Mason University, Fairfax, VA, USA
e-mail: armando@scensei.com

Randall Gray The Commonwealth Scientific and Industrial Research Organisation (CSIRO), Hobart, Australia
e-mail: Randall.Gray@csiro.au

George Grozev The Commonwealth Scientific and Industrial Research Organisation (CSIRO), Melbourne, Australia
e-mail: George.Grozev@csiro.au

Atakelty Hailu School of Agricultural and Resource Economics, University of Western Australia, 35 Stirling Highway, Perth, WA 6009, Australia
e-mail: atakelty.hailu@uwa.edu.au

S. Huet Laboratoire d'Ingénierie des Systèmes Complexes, Irstea, Aubiere, France
e-mail: sylvie.huet@irstea.fr

Marco A. Janssen School of Human Evolution and Social Change, Arizona State University, Tempe, AZ, USA
e-mail: Marco.Janssen@asu.edu

Melissa James The Commonwealth Scientific and Industrial Research Organisation (CSIRO), Melbourne, Australia
e-mail: Melissa.James@csiro.au

M. Lenormand Laboratoire d'Ingénierie des Systèmes Complexes, Irstea, Aubiere, France
e-mail: maxime.lenormand@irstea.fr

Richard Little The Commonwealth Scientific and Industrial Research Organisation (CSIRO), Hobart, Australia
e-mail: Rich.Little@csiro.au

K. Moore Faculty of Environment, Society and Design, Lincoln University, Lincoln, New Zealand
e-mail: Kevin.Moore@lincoln.ac.nz

W. Naivinit Ubon Ratchathani University, Warinchamrap, 34190 Ubon Ratchathani, Thailand
e-mail: wnaivinit@gmail.com

John Page The Commonwealth Scientific and Industrial Research Organisation (CSIRO), Melbourne, Australia
e-mail: John.Page@csiro.au

C. Le Page UPR Green, Cirad, 34000 Montpellier, France
e-mail: christophe.le_page@cirad.fr

Amanda Sahrbacher Leibniz Institute of Agricultural Development in Central and Eastern Europe, Halle, Germany
e-mail: a.sahrbacher@iamo.de

Christoph Sahrbacher Leibniz Institute of Agricultural Development in Central and Eastern Europe, Halle, Germany
e-mail: sahrbacher@iamo.de

D. Simmons Faculty of Environment, Society and Design, Lincoln University, Lincoln, New Zealand
e-mail: David.Simmons@lincoln.ac.nz

Alex Smajgl CSIRO Ecosystems Sciences, Townsville, Australia
e-mail: alex.smajgl@csiro.au

C. Smallman School of Business, University of Western Sydney, Sydney, Australia
e-mail: C.Smallman@uws.edu.au

G. Trébuil UPR Green, Cirad, 34000 Montpellier, France
e-mail: guy.trebuil@cirad.fr

J. Wilson Faculty of Environment, Society and Design, Lincoln University, Lincoln, New Zealand
e-mail: Jude.Wilson@lincoln.ac.nz

Chapter 1
Empiricism and Agent-Based Modelling

Alex Smajgl and Olivier Barreteau

1.1 The Challenge of Characterising and Parameterising Empirical Agent-Based Models

Agent-based modelling is losing its niche character and gaining wider recognition as a valuable methodology in empirical policy related situations. This growing recognition roots in the increasing demand for methods that allow integrating indicators from various disciplines across a broader systems perspective. Agent-based modelling gains its integrative strength from a combination of factors, in particular its ability

- to model explicitly cognitive processes, human decision making processes and social interactions,
- to model interactions between humans and technologies, the ecology, and physical dynamics,
- to spatially reference such cross-disciplinary interactions,
- to combine heterogeneous sources of knowledge, and
- to link variables at variable resolutions across various scales.

The increasing availability of micro-level data for humans, their behaviour and societal processes combined with the ongoing improvement of software (and computational processing power) to develop and run agent-based models have accelerated the empirical applications of this bottom-up modelling methodology. However, with this technology comes the potential for research to be deceivingly realistic, in particular when realism is a goal of the computational visualisation. Wrong model assumptions can easily be glossed over when presented in seemingly realistic visualisations. With the potential for integration and enhanced computational

A. Smajgl (✉)
CSIRO Ecosystem Sciences, Townsville, Australia
e-mail: Alex.Smajgl@csiro.au

O. Barreteau
IRSTEA, Montpellier, France
e-mail: olivier.barreteau@irstea.fr

A. Smajgl, O. Barreteau (eds.), *Empirical Agent-Based Modelling – Challenges and Solutions*, DOI 10.1007/978-1-4614-6134-0_1,

visualisation comes an amplified responsibility for robust model development and cautious model use. This introduces a set of challenges to empirical agent-based modelling, of which the approach for translating real-world data into robust model assumptions is a critical one. We refer to this process as the characterisation and parameterisation of empirical agent-based models.

In order to meet this and other challenges it seems promising to draw on diverse experiences in empirical modelling. Sharing and structuring experiences can facilitate methodological learning towards a better understanding of what solutions are promising in what context. We aim for a guide that allows newcomers to identify most effective ways, or sequences, for characterising and parameterising empirical agent-based models. We also hope that more experienced modellers find inspiration to test parameterisation sequences and, thereby contribute to an improved understanding of what technique to use in what context to improve model robustness.

Building on an earlier framework (Smajgl et al. 2011) we present in this chapter a revised characterisation and parameterisation framework and develop a decision tree to provide guidance for choosing particular methods to conduct the characterisation and parameterisation in diverse empirical situations.

1.2 Definitions

Characterization, as well as parameterization, comes prior to description of the model, for it is a part of the model design process itself. Characterisation aims at surfacing the intended model as an artefact: qualifying its contours and interfaces. Parameterisation aims at specifying the relation between the model and its target system: how suitable sources of information are incorporated. Characterization is currently embedded in formal description processes such as ODD (Grimm 2006, 2010) which gathers at the same time description of the outcome and the modelling process. In this book we want to focus on these stages because of the many approaches used but all known as "Agent based modelling".

Characterization includes first an informal step (Triebig and Klugl 2009): given the existing knowledge from theory and prior empirical experiences of the issue, what does the model intend to capture? This leads to model formulation. This characterization is progressively funnelled in the specification of a model structure: input and output spaces as well as their interfaces with the model content. This step of characterization involves explaining what should be the entities and dynamics included in the model in order to capture main features of the target system related to the issue at stake. Characterization ends with defining a model as a transformation of a situation (an element of Input space) into an element of output space, given a specific set of parameters. The following example should clarify this step: Let us assume the task is to develop a model that simulated surfers sharing waves. Furthermore we assume that observations have shown that in established communities of surfers fewer accidents occur than in communities where tourists are numerous. The modelling purpose is to understand coordination patterns among surfers. For this context the input space would be a population of surfers with knowledge of rules in use in the place where they surf and the place where they use to surf, technical level, and social network among local surfers. The output space would be a series of accidents and the quantity of good rides per surfers. Behaviour patterns compute surfers' trajectories, which are then intersected at each time step to identify occurrence of accidents.

Parameterisation aims at connecting model and target system, through giving values to the set of parameters in order to enable simulation. This means gathering knowledge from the target system to define these values, which we consider exogenous to the simulation dynamics and invariant along the simulation. The definition of these parameters not only specifies the relation between input and output of the model (agent attributes and behaviours). It also provides information on the structure of the population of the target system so that upscaling can be performed to generate a suitable artificial population. Going back to our surfer model, parameters may deal with a maximum number of surfers already riding a wave to engage in, or with classes of waves that surfers like to surf according to their experience. Parameterisation is not only a matter of giving quantitative values to parameters, but to enable running the model with a set of values. Sets of categories are particularly useful for qualitative or fuzzy approaches. After an artificial population has been generated, simulations can be performed and results can provide insights for improving the characterisation and parameterisation. A first assessment of the model at this level entails characterising relations between output indicators and input situations for sets of parameters in order to check if intended features from the target system are captured.

Characterization and parameterization come prior to calibration which is a distinct step focusing on fine tuning of parameter values so that output indicators have the observed or expected value for a given controlled input situation.

1.3 The CAP (Characterisation and Parameterisation) Framework

The goal to share experiences between modelling teams that developed empirical models in different contexts demands a generic structure, a framework that allows for a context-independent description of key steps. This work builds on a framework developed by Smajgl et al. (2011). However, this earlier framework was limited to socio-ecological contexts. Here, we expand and improve the framework to capture all human related contexts, including, for instance, socio-technological systems. The process of editing this book was a key method of testing how generic the previous framework was for a diverse set of contexts. Based on these experiences we were able to consolidate the principle steps and to add detail to the possible sequencing of optional parameterisation methods.

Figure 1.1 depicts the revised Characterisation and Parameterisation (CAP) framework. The modelling process is based on empirical and theoretical steps that initialise the research process and introduce modelling as a beneficial methodology. Subsequently, agent-based modelling is identified as an appropriate method for analysing the research question at hand.

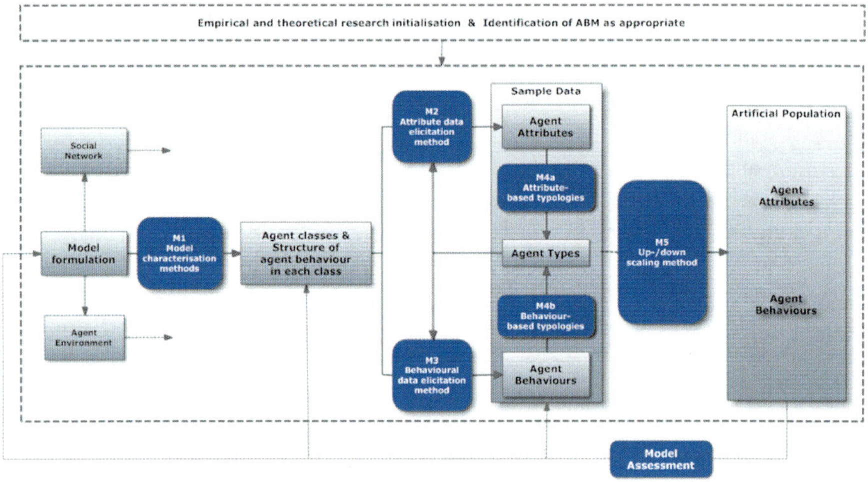

Figure 1.1 Characterisation and parameterisation (CAP) framework

Model formulation requires the definition and design of three principle components: human agents, their social network, and the environment human agents operate in. This statement implies that this work is focused on situations in which humans are modelled. This volume is focused on the process for characterising and parameterising human agents. Future volumes of this series on challenges in empirical agent-based modelling will be focused on the social network and the biophysical environment human agents operate in.

The specification of principle agent types in agent classes and principle behaviours of each agent class is a core step of the model design process. We label methods that are typically used during this step as model characterisation methods, or M1. Then, quantitative and/or qualitative data is required for the parameterisation of attributes that characterise human agents and behavioural strategies of human agents. These two sets can require different methods, depending on the modelling context. Therefore, we distinguish between methods to parameterise attributes (M2) from methods for parameterising behaviours (M3).

The next step can involve the development of explicit agent types, which can be derived from agent attributes or from agent behaviours. This simplification reduces the level of heterogeneity in the agent population but has the advantage of making the empirical implementation of the model feasible. Not all modelling processes require the development of explicit agent types. Those that develop such types have various methodological options, which we label as methods for an attribute-based typology (M4a) or a behaviour-based typology (M4b).

In most contexts data is elicited for a group of people smaller than the population in the real target system. Stepping from such a sample to a larger population requires up-scaling methods (M5). In some situations this step might involve downscaling, in particular when building on higher aggregated data. As in all previous steps, the modeller can choose between different methods (see Table 1.1) and typically faces the question "Which method most effectively improves model robustness?". Once implemented, simulation results can inform the model assessment process to iteratively improve model robustness.

Table 1.1 lists for each step in the framework the methodological options. The following Section explains these methods in more detail.

1.4 Methods at Hand

Several methods can be used to characterise and parameterise behavioural responses of humans empirically. These methods aim at capturing new knowledge about the target system and/or at transforming existing knowledge into usable information for the sake of modelling. This set of the most commonly used methods is not all-inclusive and methods are obviously not exclusive of one another. We posit that it is complete enough to build a fairly comprehensive framework:

Participant Observation Becker (1958) defines participant observation as the process in which the scientist participates in and documents the daily life of communities. This method is also used in specific collective events to which the scientist attends to or is active in, such as formal meetings of a given organisation. In this case it needs to be completed by interviews or focus groups.

Social Surveys Survey instruments consist of a list of questions, each with predefined sets of possible answers (Nichols 1991). Responses are elicited via mail, email, in person, or via phone. They are usually tailored for a large number of respondents.

Interviews While a survey comprises of mostly closed questions, interviews are normally less structured (Jupp 2006) and range from a list of open questions to unstructured dialogues. As a consequence, it is more difficult to use them for a large sample.

Knowledge Engineering Methods Close to interviews, focus groups and experimental settings, there is a whole family of ad hoc methods aiming at eliciting knowledge of people involved in a problematic situation. These elicitation techniques, such as KNeTs (Bharwani 2006), root in ethnographic approaches with the addition of devices to foster reactions of participant(s). These approaches mainly aim at grasping implicit empirical knowledge on behaviours (Becu et al. 2003).

Focus Groups While interviews are usually on an individual basis, focus groups are settings that gather a small number of people concerned with a given issue and asked to discuss it (Krueger and Casey 2000). Like interviews they provide mainly in-depth qualitative information. The main benefit compared to interviews is that participants explain more of their reasoning. The scientist launches the topics but is not part of the discussion, but for the sake of clarification (if no external facilitator). Participants have any kind of knowledge about the issue at stake from academic to

Table 1.1 Overview of methodologies relevant for the parameterisation of behavioural traits of human agents. (Smajgl et al. 2011)

M1	M2	M3	M4a ATB	M4b BT	M5
Expert knowledge	Survey	Survey	Clustering and regression	Clustering and regression	Proportional
ParticObser	Census (incl. GIS data)	Interviews	Correlation and expert knowledge	Correlation and expert knowledge	Census/GIS based assignment
Lab experiment		Field experiment	Expert knowledge	Expert knowledge	Monte Carlo
Interviews		ParticObser	Dasymetric mapping	ParticObser	
RPG		RPG			
		Time series data			
		Expert knowledge			

practical or technical. Groups can be heterogeneous or not to that regard. Preferably participants do not know each other ex ante.

Expert Knowledge This constitutes a variety of formal and informal methods for capturing the understanding of experts in a field or region and representing that knowledge in ways that it can be used in models. Approaches range from conceptual mapping by experts themselves to informal conversations or focus groups from which expert knowledge can be elicited for subsequent model building. Expert knowledge can also be employed to quantify uncertainties associated with expert judgement (Cooke and Goossens 2008). We are aware that expert comes as an increasingly fuzzy category today. Some authors like (Oliver 2012, p. 957) consider anyone is an expert, erasing thus any difference between lay and expert knowledge. We prefer to acknowledge a difference, stressing on the difference between reflexive knowledge from self practice on one hand—that we might collect through participant observation, interviews or knowledge engineering—and distant knowledge from observation, analysis or transmission from others that we consider as expert knowledge.

Census Data While surveys, interviews or focus groups are conducted with a sample of a population, census data is elicited for 100 % of a population, normally within national boundaries (Rees et al. 2002). Aggregated census data summarise responses for groups of households within enumeration districts, while disaggregated census data show household level responses.

Field or Lab Experiments While focus groups are rather open and unstructured discussions regarding a topic, an experimental setting aims at structuring interactions among a group of participants in order to observe how the change in an independent variable affects a dependent variable. Traditionally, this happened in laboratories, which allows for high degrees of control and hence clear causal links in observed outcomes. Participants are isolated and are not supposed to interact besides interactions required by the experiment. There is no room for discussion, even at the end. Participants do not need to be aware of the situation reproduced. However, the high levels of control create artificial situations, which can lead to poor applicability of results to more realistic situations (Patzer 1996). Field experiments aim for less artificialness by placing the experiment in the natural environment (Harrison and List 2004; Smajgl et al. 2009). While field experiments allow for more realistic behaviour to be observed the levels of control decrease.

Role-Playing Games Barreteau (2003) defines Role-Playing Games (RPGs) as "group settings that determine the roles or behavioural patterns of players as well as an imaginary context." Similar to experimental designs, people are given pre-defined roles and tasks, which they have to perform in interaction with other role players in a pre-defined setting. A RPG session includes a debriefing stage with participants which is crucial to consolidate knowledge acquired. This debriefing is a focus group among participants on their common experience in the game sessions and its relation to the target system. Knowledge is also acquired during the game through observation of the players behavioural patterns and their handling of game rules.

Cluster Analysis Clustering describes the grouping of subjects based on the similarity of attributes. Based on categorical data, algorithms calculate either (1) the Euclidean distance between the median and each value, or (2) between each pair of values (Rousseeuw 1987). Depending on proximity, subjects are mapped into clusters.

Dasymetric Mapping This method involves a combination of detailed spatial data with aggregated census data to create disaggregated representations of the spatial distribution of population characteristics (Mennis and Hultgren 2006). Often, for example, land-cover maps based on satellite imagery are used to distribute populations within an area such that (a) population totals within census enumeration districts are preserved and (b) their spatial locations are estimated at a much finer resolution, based on the locations of land-cover types with different population densities.

Monte Carlo Method This approach refers to "experiments with algebraic models which involve a stochastic structure" (Martin 1977). Monte Carlo runs with stochastic agent-based models allow for development of uncertainty distributions of output variables.

Regression This is a statistical method used to transform observations on a few features into knowledge on relations between these features.

Cloning In order to generate an artificial population, the modeller has to cope with the issue of upscaling. A simple method is assuming identity of agents with an agent considered as representative of a part of the population. This agent is virtually "cloned".

Proxy Data In some cases it is difficult to get empirical data as required, for example in conflict situations. The use of proxy data coming from systems supposedly comparable to the target system is then a default solution.

1.5 Options for Sequencing Methods to Characterise and Parameterise Agent-Based Models

The majority of cases require a combination of data elicitation methods. It depends on the modelling goal, availability of data, availability of funding and other contextual factors what methods the modelling team can draw on. The following decision tree aims at providing guidance to modellers for identifying what case best represents their modelling context. With each case goes a particular (and recommended) sequence of methods. Some cases outline optional additional methods for improving data quality.

Distinguishing modelling situations is critical because there is no methodological panacea across all contexts. Differentiating contexts can be done by applying many different criteria in many different ways. Figure 1.2 depicts what we identified as most effective based on the cases we considered during the editing process. In a first step we distinguish situations in which the modeller needs to model a large

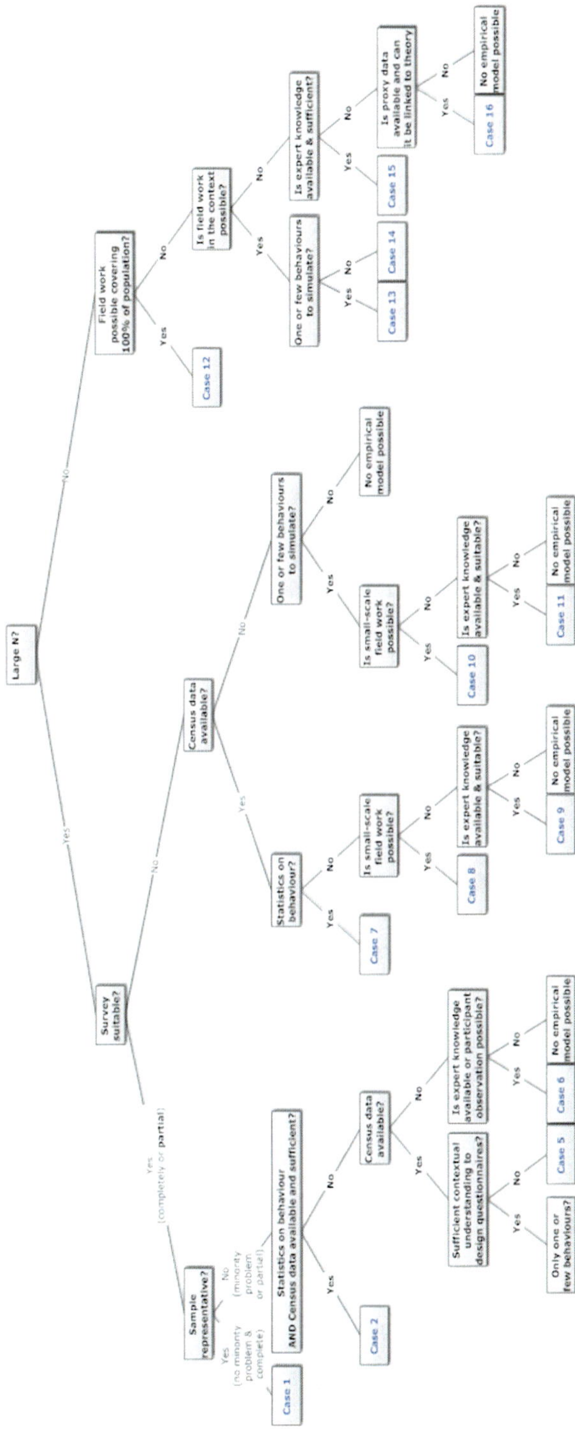

Figure 1.2 Decision tree

population of human agents from situations where this is not the case. Large populations emerged to be those that dealt with thousands and more humans. In following steps we distinguish cases (or situations) based on

- the ability to implement surveys,
- to conduct other types of field work,
- the availability of time series data on behaviour,
- the availability of census data (for attributes),
- the availability of expert knowledge
- the availability of other contextual understanding,
- the availability of proxy data, and
- the number of behaviours that need to be simulated.

Some of the resulting cases define situations in which we do not recommend the development of an empirical agent-based model as it seems unlikely that a sufficiently robust model can be developed. While situations can change as the availability of funding and skills increase or decrease it seems useful to define these 16 cases as situations in which one parameterisation sequence might work better than in another. The following explains these 16 cases in more detail before providing actual examples for empirical models within this CAP framework. Covering all 16 cases, we propose for each step of the parameterisation sequence a subset of methods.

1.6 Case 1

This case is relevant if large populations need to be modelled. The key assumption for this case is that all relevant attributes and behavioural assumptions for agents can be derived from surveys without the additional development of typologies. This implies that the proportions of responses in the sample resemble real proportions. This assumption is in particular problematic if one or more relevant behaviours are relatively rare. Capturing such minorities in a sample and performing proportional up-scaling would lead to an overrepresentation of this behaviour. Vice versa, if the behaviour is not captured in the sample the relevant behaviour is missing from the simulation. For case 1 we assume that this minority problem is not relevant. If this is a valid assumption the modeller can reproduce each sampled person or household as many times as needed to characterise and parameterise the entire agent population. This process can be referred to as cloning. It implicitly assumes that each sampled person or household is a type without explicitly developing household types.

Figure 1.3 depicts the parameterisation sequence for case 1 and the relevant methods for each step. The design of principle agent classes and principle behaviours can be informed by consulting experts, participant observation and interviewing people in the relevant system that have useful insights. Based on this design it will be possible to frame questions regarding attributes, as it should be known what attributes are relevant for simulating the system at hand. Symptomatic for this case

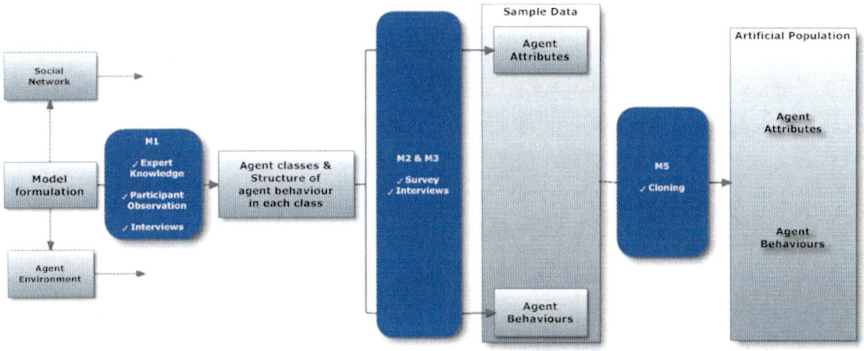

Figure 1.3 Parameterisation sequence and available methods for case 1

is that data for simulating behavioural responses can also be derived by surveys or interviews. Such questionnaires can either list questions concerning past behaviour or target intentional data, or a combination of both. These questions should reflect the changes the model user aims to specify for model simulations. The field work yields data that covers for each respondent relevant attributes and relevant behavioural responses. Based on the assumption that proportional up-scaling is adequate each response can be multiplied. Generally, parameter assumptions include range values avoiding groups of homogenous agents.

Case 1 describes a significant situation for large-N empirical agent-based modelling. Chapters 2–4 provide detailed and replicable examples for this case. Critical is the question if proportional up-scaling is an adequate procedure. In many situations it cannot be assumed that the proportions of sampled responses reflect reality. In these situations typologies need to be developed. The next three cases describe approaches frequently used for such large-N modelling situations.

1.7 Case 2

Case 2 also describes a situation in which a large population needs to be simulated. But in this case it is unlikely to obtain a representative sample by surveying parts of the population. This could be due to capacity constraints, in particular in situations with a multitude of agent behaviours requiring empirical data. In an ideal case datasets for agent attributes and behavioural responses are available. Already available datasets can be used to develop typologies for designing effective questionnaires eliciting the missing information.

Figure 1.4 provides for case 2 an overview of steps and methods for characterising and parameterising humans in empirical agent-based models. Principle agent classes and agent behaviours can be designed based on expert knowledge, participant observation or interviews. Given that time-series data and census data is

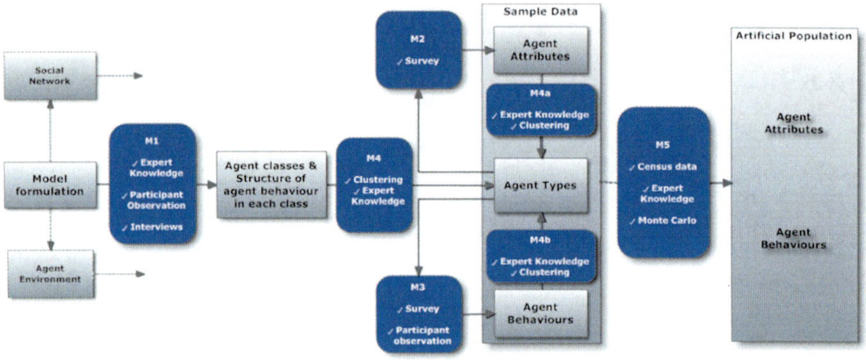

Figure 1.4 Parameterisation sequence and available methods for case 2

already available but cannot cover all parameterisation needs, the existing information can first be processed to derive first agent types. In some situations typologies have already been developed for similar contexts and can be accessed and used for the model development, as described in Chap. 5. Based on these types missing behavioural response data can be obtained by surveys or participant observation. In some situations the elicited data requires processing before mapping back into revised agent types. Then, any missing attribute data are elicited in surveys and, again, mapped back into the agent types. Depending on the information gaps the alternative sequence would be to specify first attributes and then behavioural response data.

Existing census data allows for mapping the agent types into the entire agent population. This step can involve Monte Carlo techniques. In many situations the process gains robustness if Expert Knowledge is obtained to avoid unrealistic up-scaling effects.

1.8 Case 3

Case 3 also aims at simulating large populations, with surveys an appropriate method of obtaining critical data sets for parameterising attributes and behavioural responses. However, in this case we assume that the proportions of sampled responses are not realistic. This is particularly relevant in situations where behavioural minorities are important. Changing proportions requires a means of mapping responses in more realistic proportions into the entire population to be simulated.

Figure 1.5 provides for case 3 an overview for steps and methods for characterising and parameterising humans in empirical agent-based models. Principle agent classes and agent behaviours can be designed based on expert knowledge, participant observation or interviews. The next step involves eliciting data on behavioural responses the simulation requires. This information can be obtained by surveying the population or by participant observation. The larger the population and the more types of agent behaviours requiring parameterisation the more effective surveys are.

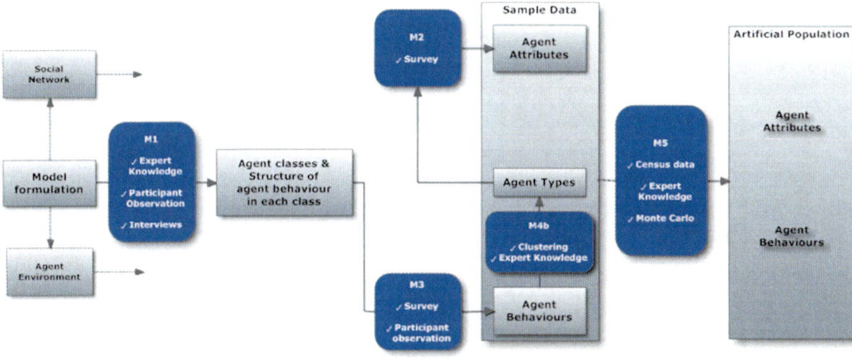

Figure 1.5 Parameterisation sequence and available methods for case 3

Assuming that no representative sample can be obtained, in particular regarding the unrealistic frequency of some responses, the data needs to be processed for identifying agent types. Typically, statistical methods, such as cluster analysis are applied, but also more qualitative approaches (like expert consultation) could lead to effective agent types. It is assumed that each type represents a specific behavioural response. In a next step, data for the parameterisation of attributes is obtained by a survey. Typically, such a survey targets core representatives of each behavioural type by asking upfront those questions that map people clearly into each behavioural type. This can be performed by using the behavioural variables with the highest discriminatory power, i.e. principle components. If during the field work persons can be identified as core representatives the entire survey is performed with them; otherwise, the person is not surveyed and the surveyor moves to the next potential respondent.

Once all necessary data for parameterising attributes and behavioural responses are elicited the data need to be mapped into the entire population. Due to the initial assumption regarding sample representativeness responses cannot be proportionally up-scaled, as done in case 1. Instead, the sample is mapped into some sort of disaggregated census, such as households level census data or GIS data for the level of human entities. This requires a robust overlap between the (pre-existing) census data and the elicited attribute. The combination of attributes must point at a specific behavioural type. Mapping each behavioural type into census data provides the required parameters for the entire agent population. Additional methods such as expert knowledge and the application of Monte-Carlo techniques can add to the robustness of the parameterisation sequence.

1.9 Case 4

Similar to the two previous cases the large-N population can be surveyed but representativeness of the sample is problematic. In contrast to Case 2 this situation does not allow for proportional up-scaling. Similar to case 3 dis-proportional up-scaling

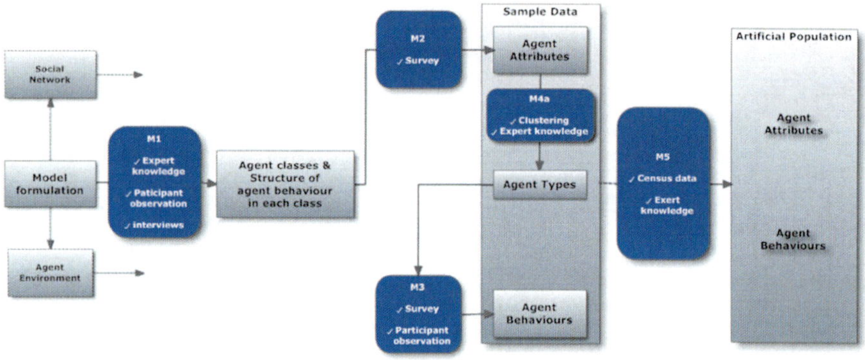

Figure 1.6 Parameterisation sequence and available methods for case 4

is performed based on agent types. However, in this case agent types are derived from attribute data and not, as in case 3, from behavioural data.

Figure 1.6 presents the parameterisation sequence for Case 4. The principle systems understanding required for designing agent classes and types of agent behaviours can be obtained from experts, through participant observation or interviews. Then, for each agent class necessary attributes are elicited in surveys. Turning first to attributes seems to contradict the essential importance of behavioural aspects, the core dimension of agent-based modelling. However, evidence suggests that people follow the same activity (i.e. livelihood) for the same reason (i.e. motivation) given similar constraints (i.e. education) would respond in a similar way to the same change (Trébuil et al. 1997; Marshall and Smajgl 2013; Barnaud 2005). Based on this assumption attribute data can be obtained covering relevant activities, motivation and constraints. Then, the data can be used in a clustering approach to develop agent types. Involvement of experts in the process, to confirm the statistical results, is recommended. For each agent type behavioural response data can be obtained in surveys or by participant observation. It is important to start the questionnaire with questions that allow the identification of the agent type the respondent is likely to represent. Most effective are the variables with the highest discriminatory power in the clustering. Only those respondents that clearly represent a particular type are surveyed. There is a clear need to find multiple respondents for each type. Chapter 7 provides an example for this process.

Once attribute data and behavioural response data are obtained and linked to agent types, these agent types need to be mapped into the entire agent population. This disproportional up-scaling is guided by census data—matching attributes of behavioural types to census information. This can be problematic if only a few attributes from step M2 match the Census. Designing the questionnaire with the up-scaling in mind allows for effectively preparing this final step.

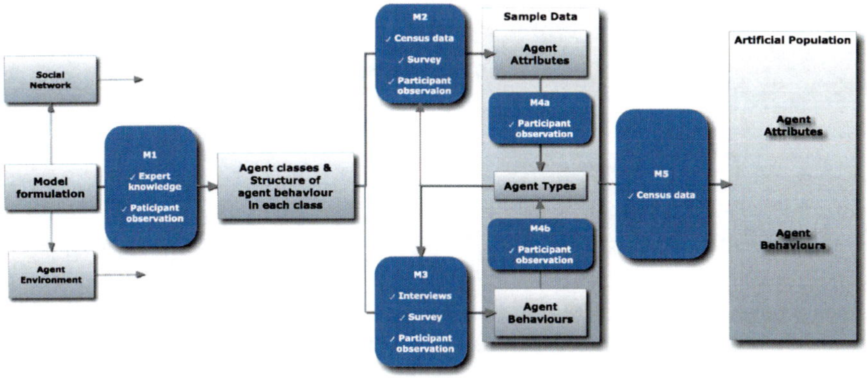

Figure 1.7 Parameterisation sequence and available methods for case 5

1.10 Case 5

Case 5 describes a difficult situation, in which a model needs to be developed for a large population, without sufficient data and without the contextual understanding to perform or design the characterisation and parameterisation process. However, we assume that for this case census data is available.

In such a situation expert knowledge or participant observation can provide broad systems understanding (M1) to characterise the model as the first step of model design. Then, in an iterative approach, sample data can be elicited to provide the necessary depth in contextual understanding (see Fig. 1.7 for steps M2 and M3). Then, participant observation allows for developing agent types based on attribute data (census or surveys) and behavioural data (interviews or role playing games). These types can be use in the next round M2 and M3 to elicit more robust data for agent attributes and behaviours. The development of types with their respective data for attributes and behaviour might have to be repeated if agent types prove to be blurred. The final up-scaling could be guided by census data.

1.11 Case 6

Case 6 deals again with large-N populations and faces a situation in which a survey is in principle suitable for eliciting relevant data for agent attributes. Similar to Cases 3–5 response proportions in the sample are not considered to be realistic, which requires an up-scaling approach that changes sample proportions, or disproportional up-scaling. However, in this case no Census data is available, which creates a major problem for disproportional up-scaling. We assume for Case 6 that adequate expert knowledge is available to perform the up-scaling.

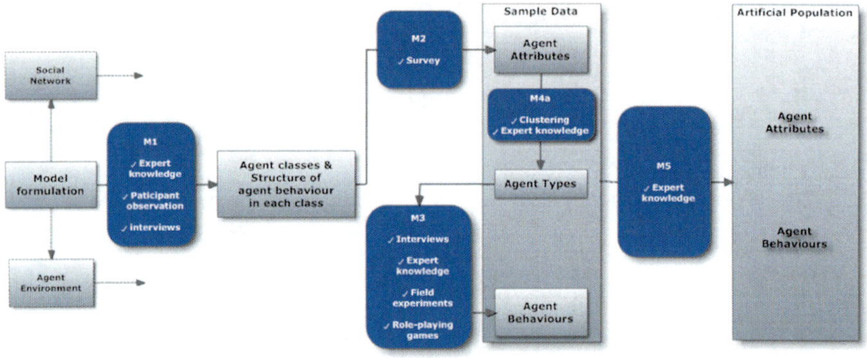

Figure 1.8 Parameterisation sequence and available methods for case 6

Figure 1.8 depicts the parameterisation sequence for Case 6. Principle systems understanding for specifying agent classes and types of agent behaviours can be derived from expert knowledge, participant observation or by conducting interviews. Based on the agent classes data requirements for agent attributes can be specified and elicited by conducting a survey. Once the survey data has been obtained it needs processing to identify agent types. Then, for each type behavioural response data needs to be obtained. This data can be elicited by conducting interviews, consulting experts, field experiments, or by role-playing games.

Once all necessary data for attributes and behavioural responses are obtained for all agent types the up-scaling is performed based on expert knowledge. Experts specify the proportions for each types and, if necessary their location. This sequence is similar to the technique we recommend for case 4. However, the absence of census data forces the modeller to replace a data-based mapping exercise by expert based approximations. To achieve acceptable model robustness requires even more diligence during the implementation of prior steps (M1-M4) and some sensitivity runs to test how variations in the proportions of behavioural types impact on simulation results. It is likely that in many instances of this case it is not possible to develop a robust empirical model as for several other situations identified in Fig. 1.2.

1.12 Case 7

Case 7 describes a situation where the population cannot readily be surveyed, for instance due to funding limitations. However, in this case census data and time series data are available to inform the relevant agent attributes and behaviours. In contrast to Case 2 no additional survey is necessary (nor would it be possible under the assumed conditions).

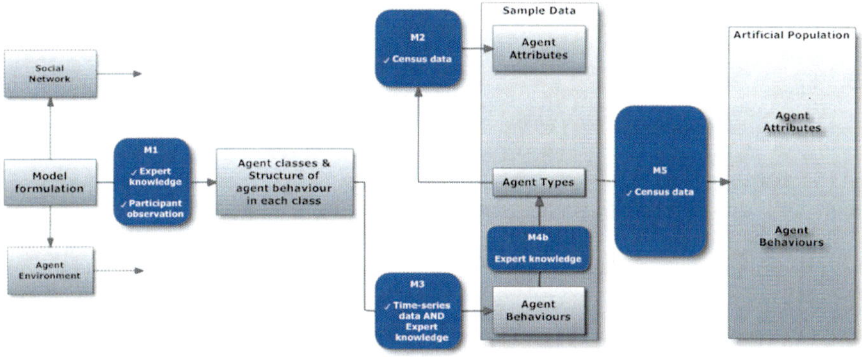

Figure 1.9 Parameterisation sequence and available methods for case 7

Figure 1.9 depicts the parameterisation sequence for Case 7, in which the general systems understanding is obtained from experts or participant observation to specify agent classes and principle agent behaviours. For each relevant behaviour time series data is obtained, statistically analysed and discussed with experts to specify and parameterise behavioural response rules. This step provides types of agent behaviours and might require many iterations between M3 and M4b. Symptomatic for Case 7 is that agent attributes can be parameterised by Census data (M2). The Census data is utilised to map behavioural response rules into the entire agent population (M5).

1.13 Case 8

Case 8 describes a case in which a large population needs to be simulated without the ability to conduct a large scale survey. However, Census data is available to inform attribute parameters and guide the up-scaling. And while no behavioural data is available small-scale field work is possible.

Figure 1.10 shows that agent classes and principle behaviours can be specified based on expert knowledge, participant observation, interviews, experiments, and role-playing games. Given the availability of Census data and the lack of behavioural data the next step in the parameterisation sequence targets attribute data. Census data is employed to parameterise agent attributes before conducting a cluster analysis to derive agent types. Alternatively, or better still in support of, the statistical procedure experts can be consulted to develop agent types. Then, behavioural data is obtained for each agent type by either conducting interviews, or consulting experts, or conducting field experiments or role-playing games, or by participant observation. Naturally, a combination of these methods can increase the robustness of the parameterisation. The sample is then up-scaled by mapping behavioural specifications according to matching attributes into Census data.

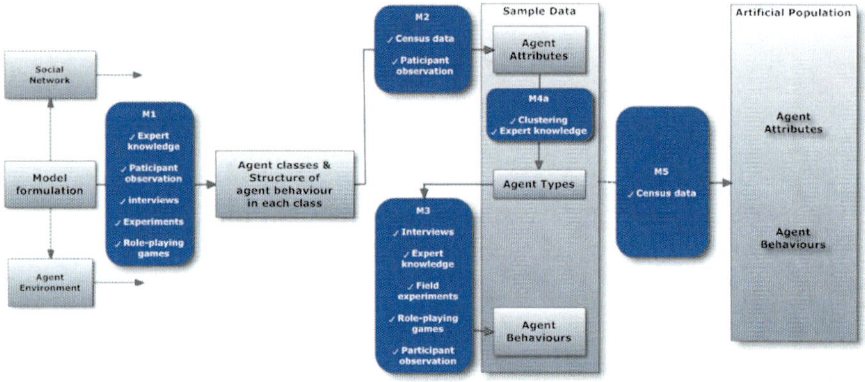

Figure 1.10 Parameterisation sequence and available methods for Case 8

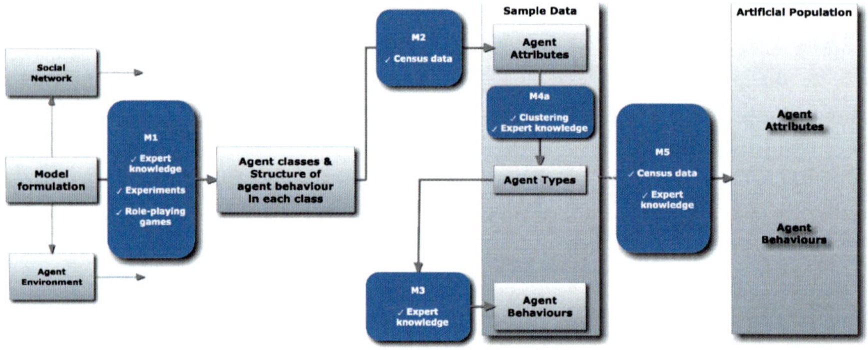

Figure 1.11 Parameterisation sequence and available methods for case 9

1.14 Case 9

Case 9 assumes that a large population needs to be simulated while no primary data can be elicited in the field. However, Census data is available and adequate expert knowledge is available.

Figure 1.11 depicts the parameterisation sequence for case 9. This approach aims for effectively combining expert knowledge and Census data. Agent classes and relevant behaviours are identified in consultation of adequate experts. Additionally, it seems useful to conduct lab-experiments or role-playing games, in particular to identify relevant agent behaviours. Then, attribute data is obtained from Census data and a cluster analysis is performed to derive agent types. It improves model robustness if experts are consulted in specifying agent types, in particular to add context to the statistical results. Then, experts are consulted to determine for each agent type behavioural rules and parameter values. The results are up-scaled to inform the

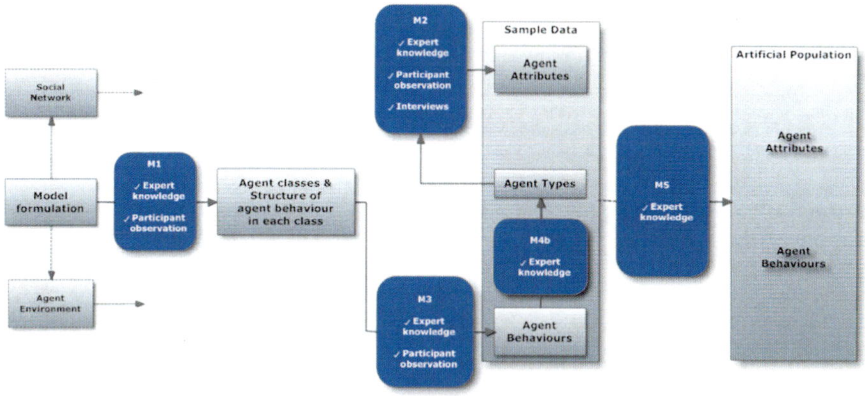

Figure 1.12 Parameterisation sequence and available methods for case 10

entire population by utilising the available Census data. Involving experts during this exercise can be effective, in particular for designing the procedure for mapping types and their behaviours into the (attribute-focused) census.

1.15 Case 10

Case 10 describes a model development that builds largely on expert knowledge. In this case a large population needs to be simulated without the possibility to conduct a large-scale survey and without access to census data or behavioural data. However, limited field work is possible to elicit contextual data first hand. Given such methodological constraints empirical agent-based modelling seems only adequate if not more than a few behaviours are simulated and if adequate expert knowledge can be employed.

Figure 1.12 shows for Case 10 that the principle systems understanding for specifying agent classes and principle behaviours is obtained from expert knowledge or by participant observation. The same methods are employed to elicit behavioural response data, best in combination. Analysing the behavioural data with experts can allow the development of agent types. Then, a combination of expert knowledge, participant observation and small-scale interviews can be employed to obtain data for parameterising agent attributes. The sample data can then be up-scaled in consultation with experts.

The lack of available data and resources for obtaining primary data is a critical problem in empirical agent-based modelling. An expert-driven parameterisation process with some supplementing small-scale field work can only be adequate for focused models that aim at simulating one or two behaviours. This approach is likely to be too limiting for models that aim for more complex simulations as robust model specifications cannot be derived.

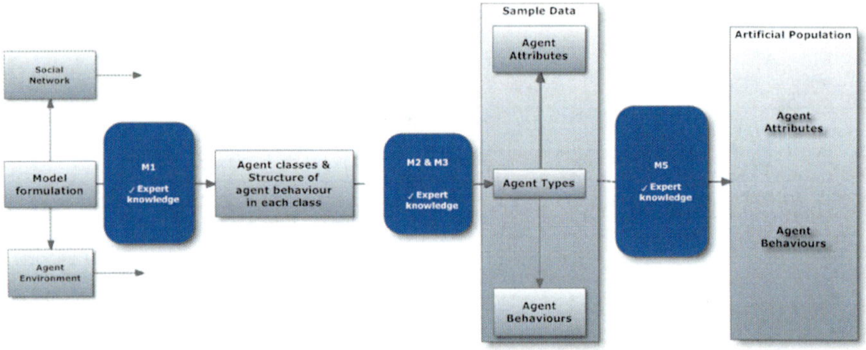

Figure 1.13 Parameterisation sequence and available methods for case 11

1.16 Case 11

Case 11 describes a situation even further constrained than Case 10: while a large population needs to be simulated there is no possibility to conduct any field work and no access to Census data. Only expert knowledge can be obtained.

Figure 1.13 shows the purely expert-driven parameterisation sequence. Principle systems understanding is developed with experts, specifying agent classes and behaviours. Then, agent types are determined based on expert knowledge, specifying behavioural responses and agent attributes. In a final step, experts specify how such agent types map into the relevant population, determining relevant aspects such as the ratios of agent types and their geographical location, if relevant.

Clearly, in such data-constrained situations the development of an empirical agent-based model is very challenging. The actual context will determine if the development of a robust empirical agent-based model is at all possible. In particular, the number of behaviours required for the simulation is critical; the more different behaviours agents need to perform the less agent-based modelling seems an adequate methodology.

1.17 Case 12

Case 12 involves relatively small populations and assumes a situation in which 100 % of the population that needs to be simulated can be accessed by field work. This makes up-scaling obsolete.

Figure 1.14 provides an overview of the parameterisation sequence for Case 12. Due to the small size of the population to be modelled agent classes and principle behaviours can be identified based on expert knowledge, participant observation, interviews, experiments, or role-playing games. Agent attributes can be elicited in

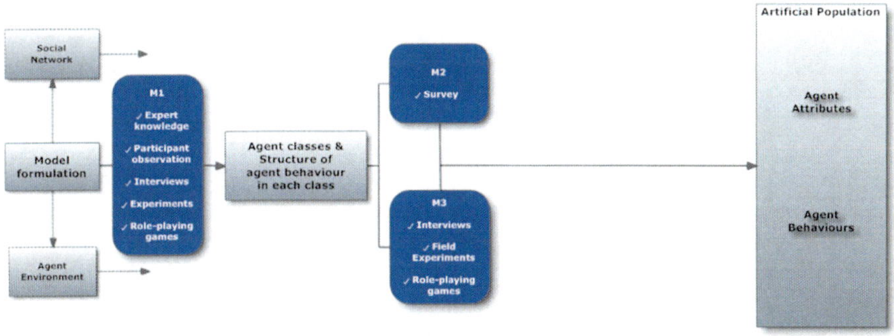

Figure 1.14 Parameterisation sequence and available methods for case 12

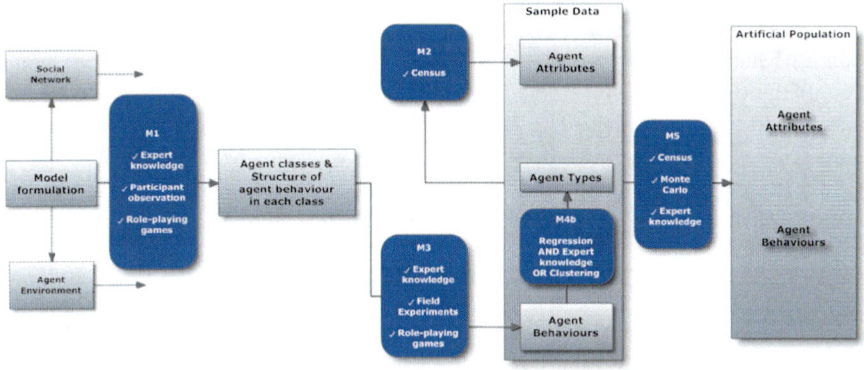

Figure 1.15 Parameterisation sequence and available methods for case 13

surveys and behavioural response data can be obtained through interviews, field experiments or role-playing games. These two steps could inform each other requiring an iterative approach revising, differentiating or expanding attribute data or behavioural response data.

1.18 Case 13

In cases with small populations and the ability to conduct field work accessing parts of the population at two cases have to be distinguished. In Case 13 we assume that only a few behaviours need to be simulated, while Case 14 covers the situation with many diverse behaviours.

Figure 1.15 depicts the parameterisation sequence for Case 13. Agent classes can be identified using expert knowledge, participant observation or role playing games. Expert knowledge, Field experiments or role-playing games lead to the relevant

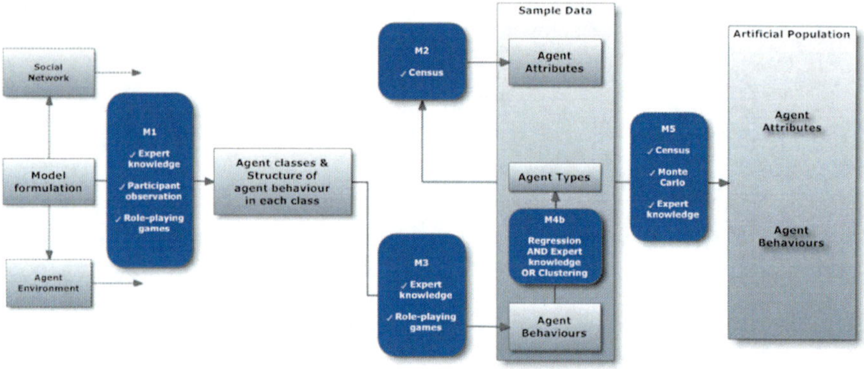

Figure 1.16 Parameterisation sequence and available methods for case 14

data for behavioural responses. This data can be used to develop agent types using a combination of regression and expert knowledge. For each type attribute data can be obtained by Census data. The sample of data for agent attributes and behaviours is up-scaled by employing Census data, Monte Carlo runs, or expert knowledge to parameterise the entire population.

Case 13 is one of the most frequent cases in the domain of small-N models (Barnaud et al. 2010; Dung et al. 2009; Mathevet et al. 2003; Smajgl et al. 2009). Chapter 11 provides a replicable example for Case 13.

1.19 Case 14

Similar to Case 13 but for situations in which many different agent behaviours need to be implemented, Case 14 follows a very similar parameterisation sequence to Case 13. The only difference is that field experiments seem less suitable for obtaining data on behavioural responses. The reason being that with increasing number of behaviours the experimental design becomes increasingly difficult, in particular if behaviours are not independent (Smajgl et al. 2008).

Figure 1.16 depicts that expert knowledge, participant observation and role-playing games can be employed to specify agent classes and principle agent behaviours in this case. The actual parameterisation starts with behavioural responses by consulting experts or by conducting role-playing games. Regression analysis combined with expert knowledge allow for developing agent types. Then, Census data is used to parameterise attribute data. The final up-scaling is performed by utilising again the Census data and mapping type-specific behavioural response rules into the entire agent population.

1.20 Case 15

Some empirical situations do not allow for any field work, due to war, violence, lack of access, or lack of funding. We specify such situations as case 15 if experts are available to provide adequate and sufficient contextual understanding to characterise and parameterise an agent-based model. Expert knowledge becomes thereby the key method for all steps M1-M5. Three variations exist for case 15 and it depends on the size of the population which of these variations is most effective. First, the process can resemble case 1 (see Figure 1.3) with only expert knowledge available for all steps. This approach is recommended if the population is too large to be parameterised one by one. Second, in situations in which all agents can be parameterised one by one, agents can be parameterised as depicted for case 12 (Figure 1.14). This is likely tot be the most robust approach. Third, assuming the second approach is not possible and that the first approach is insufficiently robust, we recommend an approach similar to case 11 (see Figure 1.13) with only expert knowledge available for all steps. This approach would develop agent types first and specify then attributes and behaviours for each agent type. In a final step, experts advise the modeller in what proportions agent types need to be placed where in the model. It is advised to run this parameterisation approach in a few iterations to establish robust model assumptions.

1.21 Case 16

Case 16 describes an extreme case of small population models as no field work is possible and no experts can be consulted. The development of an empirical agent-based model is only justified if proxy data can be legitimately utilised for the characterisation and parameterisation of human agents. Proxy data describes data from a different place but with a context similar to the one that needs to be simulated.

Figure 1.17 shows that agent classes and relevant behaviours can be specified by proxy data. Approximating across contexts can be difficult and it contributes to the methodological robustness if key aspects are tested in lab-experiments or role-playing games. Once agent classes and principle behaviours are specified proxy data is used to derive attribute data and behavioural data. As before, the approach gains robustness by conducting lab-experiments or role-playing games for behavioural data. Once both datasets are compiled statistical methods are employed to specify agent types, in particular regression analysis for behavioural response data and cluster analysis for attribute data. In such a data limited situation up-scaling is challenging but can be possible by employing proxy data. It is advisable to analyse at the beginning if the available proxy data allows the modeller to map sample data into the entire agent population.

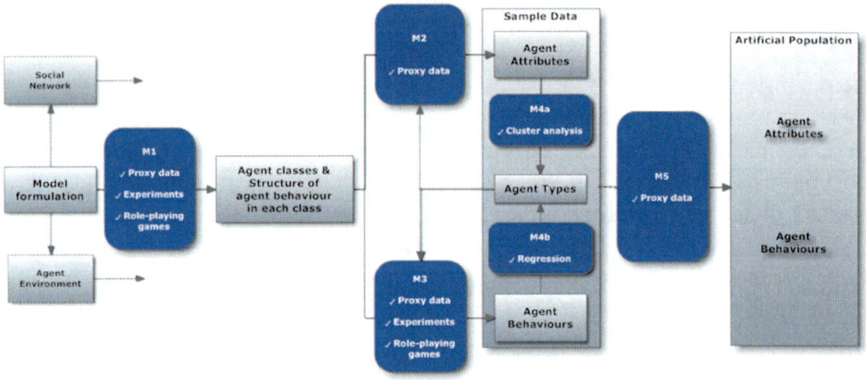

Figure 1.17 Parameterisation sequence and available methods for case 16

1.22 From Challenges to Solutions

This volume aims to provide a generic framework that allows for a structured and unambiguous description of the characterisation and parameterisation approach a modeller implemented. Additionally, cases were defined to distinguish particular modelling situations, which might require different methods for robust model characterisation and parameterisation. Both combined allows for a comparison of how effectively different methods perform in similar contexts. We hope that this comparative work will contribute to an improved methodological robustness of agent-based modelling in empirical situations.

We hope that distinguishing modelling situations in separate cases with their particular sequence of recommended methods can also be read like a cooking book with recipes newcomers can easily follow. As the above is based on abstract conceptualisations of modelling situations and scientific methods we see the need to provide real examples for the most prominent cases. These examples are provided in Chaps. 2–12 and describe actual characterisation and parameterisation processes with sufficient detail to allow the reader to replicate the same technique. But again, if the modelling situation differs to the one faced by the authors of the following chapters, the particular sequence might not be practical.

During the editing process of this book it became obvious that empirical agent-based models have not been developed for all theoretically possible cases. Therefore, this volume does not provide an example for all cases identified in Fig. 1.2. However, the most frequent cases of empirical agent-based modelling seems to be cases 1 and 7 for large populations and case 13 for systems with small human populations. Within each of these cases different methods can be combined, which allows for the comparative discussion this volume aims to initiate.

The final chapter will endeavour to initiate a comparative discussion after addressing critical questions regarding the efficacy and robustness of the framework and the decision tree. We understand this as a first step and believe that the framework is likely to need further revision to structure a wider range of empirical agent-based models. Ultimately, the *raison d'être* of this work is a robust, defensible and widely accepted means by which agent-based models are implemented in empirical situations and, thereby, to contribute to the advance of agent-based modelling itself.

References

Barnaud C (2005) Agriculture paysanne du Nord-Est thailandais. Les Indes Savantes, Paris

Barnaud C, Van Paassen A, Trébuil, G, Promburom T, Bousquet F (2010) Dealing with power games in a companion modelling process: lessons from community water management in Thailand Highlands. J Agri Educ Ext 16:55–74

Barreteau O (2003) The joint use of role-playing games and models regarding negotiation processes: characterization of associations. JASSS—J Art Soc Social Simulat 6. http://jasss.soc.surrey.ac.uk/6/2/3.html%3E

Barreteau O, LePage C, D'Aquino P (2003) Role-playing games, models and negotiation processes. JASSS—J Art Soc Social Simulat 6.

Becker HS (1958) Problems with inference and proof in participant observation. Am Sociolog Rev 23:652–660

Becu N, Perez P, Walker A, Barreteau O, Page CL (2003) Agent based simulation of a small catchment water management in northern Thailand: description of the CATCHSCAPE model. Ecol Model 170:319–331

Bharwani S (2006) Understanding complex behavior and decision making using ethnographic knowledge elicitation tools (KnETs). Soc Sci Comp Rev 24(1):78–105

Cooke RM, Goossens LLHJ (2008) TU delft expert judgment data base. Reliab Eng Syst Saf 93:657–674

Dung LC, Hoanh CT, Le Page C, Bousquet F, Gajaseni N (2009) Facilitating dialogue between aquaculture and agriculture: lessons from role-playing games with farmers in the Mekong Delta, Vietnam. Water Policy 11:80–93

Grimm V, Berger U, Bastiansen F, Eliassen S, Ginot V, Giske J, Goss-Custard J, Grand T, Heinz SK, Huse G, Huth A, Jepsen JU, Jørgensen C, Mooij WM, Müller B, Pe'er G, Piou C, Railsback SF, Robbins AM, Robbins MM, Rossmanith E, Rüger N, Strand E, Souissi S, Stillman RA, Vabø R, Visser U, DeAngelis DL (2006) A standard protocol for describing individual-based and agent-based models. Ecol Model 198:115–126

Grimm V, Berger U, DeAngelis DL, Polhill JG, Giske J, Railsback SF (2010) The ODD protocol: a review and first update. Ecol Model 221:2760–2768

Harrison G, List J (2004) Field experiments. J Eco Literature 42:1009–1055

Jupp V (2006) The sage dictionary of social research methods. Sage, London

Krueger R, Casey MA (2000) Focus groups: a practical guide for applied research, 3rd edn. Sage, Thousand Oaks

Marshall NA, Smajgl A (2013) Understanding variability in adaptive capacity on rangelands. Rangeland Ecology and Management 66:88–94.

Martin LR (1977) Survey of agriculture economics literature: quantitative methods in agricultural economics, 1940–1970s. University of Minnesota Press, Minneapolis

Mathevet R, Bousquet F, Le Page C, Antona M (2003) Agent-based simulations of interactions between duck population, farming decisions and leasing of hunting rights in the Camargue (Southern France). Ecol Model 165:107–126

Mennis J, Hultgren T (2006) Intelligent dasymetric mapping and its application to areal interpolation. Cartogr Geogr Inf Sci 33:179–194

Nichols P (1991) Social survey methods: a fieldguide for development workers. Oxfam, Oxford

Oliver DM, Fish RD et al (2012) Valuing local knowledge as a source of expert data: farmer engagement and the design of decision support systems. Environ Model Softw 36:76–85

Patzer GL (1996) Experiment-research methodology in marketing: types and applications. Quorum Books, Westport

Rees P, Martin D, Williamson P (2002) The census data system. Wiley, New York

Rousseeuw PJ (1987) Silhouettes: a graphical aid to the interpretation and validation of cluster analysis. J Comput Appl Mathematics 20:53–65

Smajgl A, Heckbert S, Ward J, Straton A (2009) Simulating impacts of water trading in an institutional perspective. Environ Model Soft 24:191–201

Smajgl A, Izquierdo L, Huigen M (2008) Modelling endogenous rule changes in an institutional context: The ADICO sequence. Advan Complex Syst 2:199–215

Smajgl A, Brown DG et al (2011) Empirical characterisation of agent behaviours in socio-ecological systems. Environ Model Softw 26(7):837–844

Trébuil G, Kam SP et al (1997) Systems diagnoses at field, farm and watershed levels in diversifying upland agroecosystems: towards comprehensive solutions to farmers' problems. In: Teng PS, Kropff MJ, ten Berge HFM, Dent JB, Lansigan FP, van Laar HH (eds) Applications of systems approaches at the farm and regional levels. Kluwer Academic, The Hague, 99–114

Triebig C, Klugl F (2009) Elements of a documentation framework for agent-based simulation models. Cybern Syst 40(5):441–474

Chapter 2
A Case Study on Characterising and Parameterising an Agent-Based Integrated Model of Recreational Fishing and Coral Reef Ecosystem Dynamics

Lei Gao and Atakelty Hailu

2.1 Introduction

Managing recreational fishing is among the most difficult natural resource management problems. The complex nature of the impacts caused by management changes makes it difficult to identify the full range of ecological and socio-economic effects. It is difficult to distinguish approaches that are effective from those that are not. For example, the evaluation of area closure strategies needs to incorporate the relationships among stock dynamics, angler responses and consequent changes in the geographical distribution of fishing efforts. Empirically-based tools are needed to predict responses to, and outcomes from, management decision that affect fish stocks and fishing benefits. To address this, an integrated agent-based model (ABM) of recreational fishing and a coral reef system is developed to evaluate ecological and economic impacts.

Recreational fishing is an individual based activity, with individuals making decisions on fishing site based on their own preference, knowledge and expectations. In this model, the behaviour of angler agents is represented by empirically based Random Utility Models (RUMs) (McFadden 1974; Schuhmann and Schwabe 2004) that rationalize choices on the basis of attributes of the individuals, the characteristics of alternative sites and recreational experience. With this approach, it is possible not only to simulate fishing behaviour but also construct welfare estimates at the individual level (i.e. for each angler), allowing resource managers and policy makers to assess the impacts of management change on different segments of society. Further, these welfare estimates can be aggregated up to the population level

L. Gao (✉)
CSIRO Land and Water, PMB 2, Glen Osmond, Adelaide, SA 5064, Australia
e-mail: dr.leigao@gmail.com

A. Hailu
School of Agricultural and Resource Economics, University of Western Australia,
35 Stirling Highway, Perth, WA 6009, Australia
e-mail: atakelty.hailu@uwa.edu.au

A. Smajgl, O. Barreteau (eds.), *Empirical Agent-Based Modelling – Challenges and Solutions*, DOI 10.1007/978-1-4614-6134-0_2,

(i.e. for all anglers) for use in cost-benefit analysis and the evaluation of changes in recreational management. The model makes it possible to undertake "what-if" scenario analyses and allows researchers and managers to better understand the wide range of economic and environmental implications of management strategies.

While ABMs have been used to study different natural resource management problems, there have been very few studies that have employed behavioural models that are grounded on empirically estimated choice models. In addition, our model couples these behavioural models with a coral reef ecosystem model that simulates the interactions among algae, corals, herbivorous and piscivorous fish. This model is incorporated into the ABM-RUM framework as a means of attributing environmental changes to recreational fishing sites.

The integrated ABM has been used to undertake demonstrative simulations and these results have been reported in several conference proceedings and journal articles (Gao et al. 2010; Gao and Hailu 2010b, 2011a, b, 2012, 2013). The two-way interaction between fishing site choices (human behaviour) and ecosystem dynamics is complex. The implications of this complexity are that it is difficult to determine the socioeconomic and biological outcomes of a management change or the relative performance of alternative management strategies without the benefit of integrated modelling. For example, Gao and Hailu (2011b) illustrate this by simulating the effects of three alternative site management strategies: a baseline strategy where no fishing sites are closed; a 2 month closure of a site; and a 6 month closure of a site. The alternative strategies are compared in terms of fish biomass and angler economic welfare outcome streams obtained over time. Further, these comparisons are done for two different fishing pressure environments: a low level (or baseline) fishing pressure level and a high fishing pressure level. Among the model's surprising conclusions is that, under low fishing pressure in a coral reef ecosystem, closing fishing areas for 2 months instead of 6 months can result in larger fish stocks and better fishing opportunities. These observations highlight the need for the use of simulation platforms to track complex outcomes and to help managers and other stakeholder explore conservation and economic tradeoffs implied by alternative resource management choices. Further details are provided in Gao and Hailu (2011b).

2.2 Model Description Based on ODD

In this section, we use a model documentation protocol, ODD (Grimm et al. 2010), to describe the integrated ABM.

2.2.1 Purpose

As indicated above, the purpose of the integrated ABM model is simulating recreational fishing and reef ecosystem dynamics. It allows resource managers to evalu-

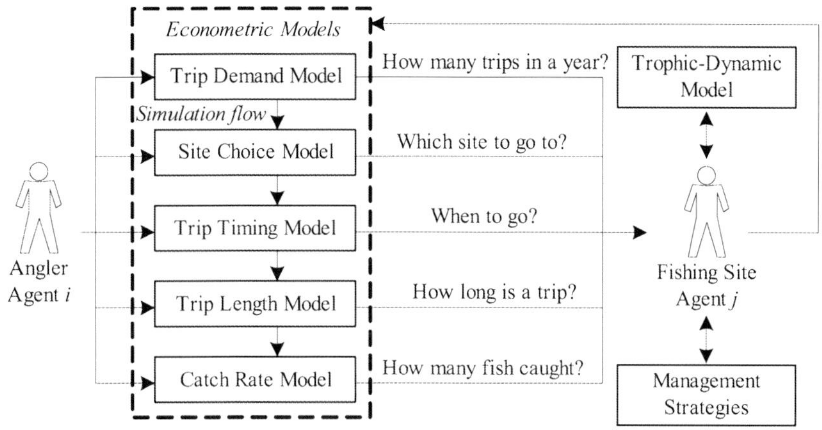

Figure 2.1 The integrated ABM. (Source: Modified from Gao and Hailu 2011b)

ate both the welfare and biophysical impacts of proposed or potential changes in management.

2.2.2 State Variables and Scales

The ABM combines a host of econometric models with a trophic-dynamic model of a coral reef ecosystem. A schematic diagram of the major components in the integrated ABM is presented in Fig. 2.1.

This ABM model is the first model to combine econometrically estimated models of agent behaviour with a biophysical model of coral reefs. Recreational anglers and fishing sites are all modelled as agents. As shown in Fig. 2.1, five econometric models (trip demand model, trip timing model, trip length model, site choice model, and catch rate model) underpin the decision-making process on which a recreational angler's behaviour is structured. These models predict, respectively, the number of recreational trips taken in a year, the timing of a trip in a year, the length or duration of a trip, the choice of recreational site in any one trip, and the agent's expected fish catch for any given site. The coral reef ecosystem model describes interactions among four components in a coral reef environment, namely, algal growth, coral cover, herbivore fish and piscivore fish. These constitute the simulation platform and are described in the "sub models" section.

2.2.3 Process Overview and Scheduling

Its main process can be summarized briefly as follows. For each angler, the simulation system generates a schedule of fishing trips and fishing site choices using behavioural models that are econometrically estimated using observed data. The angler's fishing schedule and choice of sites depends on personal or angler attributes, his/her fish catch expectations (site by site) and the set of fishing sites available as well as the nature of those sites. If an angler agent is fishing on a particular day, then that angler will make a trip to his/her chosen site for that trip. Fish catch for angler at the chosen site will depend on the fish stocks at the site as well as on the angler's demographic or personal attributes. This catch is again determined by an empirically estimated model as described later in the section on "sub models". Thus a fishing trip is affected by the conditions on the site. Fishing activities in turn affect fish stocks (and the rest of the ecosystem) at a site. Each fishing site is a coral reef ecosystem (modelled as a trophic-dynamic system) where fish stocks, coral cover, and algal cover are affected by the fishing behaviours of angler agents. These effects will then feedback into fishing site choices through impacts on catch rates. That is, there is effect going both ways, from the biological to the economic and back. And these effects are complex. For example, when a site is closed, the choice available to the angler is limited. This redistributes fishing effort and has the potential to affect conditions in other sites. The consequence of this redistribution will have further effects on fishing behaviour, etc. The integrated model is used to tease out these effects in a consistent manner so that the socioeconomic and ecological consequences of changes in management strategies are easier to evaluate.

2.2.4 Initialization

The Initialization in the integrated ABM involves specification of fishing site agents and angler agents. A number of fishing site agents are created with corresponding data on fish stocks, coral cover, and algal cover, which are based on information about the study areas and collected field data. A scaling method is used to initialise a population of anglers with demographic characteristics and recreational fishing attributes. In addition, the simulator works on a daily basis, meaning that one modelling time step is equivalent to a 24-h day in reality.

2.2.5 Input

Further inputs are required once the model is initialized. The characteristic information of a calendar day, such as whether the day is a weekend, a public or a school holiday, are incorporated into the ABM and affect the angler agent's make decisions on trip timing and trip length. Information on the species of fish caught and the distance between sites is used as input to enable a more accurate modelling of fishing costs.

2.2.6 Sub Models

In this section, we describe in more detail the sub models, starting with the econometric models that form the basis of our angler behaviour characterisation. The section concludes with a brief description of the coral reef ecosystem model that is coupled with the human behavioural model to provide an integrated model that takes into account the two-way interaction between fishing behaviour and biophysical outcomes.

The *trip demand model* predicts the actual number of trips taken by an angler (in a year) as a Poisson process (Raguragavan et al. 2013). The logarithm of number of trips in a year λ_i is specified as a function of the expected maximum utility from a fishing trip, known as "inclusive value" (IV) in the economics literature, and a set of socio-economic characteristics of the angler. In particular, the model is specified as in Eq. (2.1).

$$\ln \lambda_i = \beta_0 + \beta_1 \cdot IV_i + \sum_m \beta_m \gamma_m \tag{2.1}$$

where γ_m represents m-th individual characteristic, such as age, education, employment, etc. β_0, β_1, and β_m are regression coefficients. The IV_i variable, which is a measure of the expected maximum utility from a set of choices, is routinely used to evaluate environmental changes in the non-market literature (McFadden 1974). It is calculated from site utility values using the formula in Eq. (2.2).

$$IV_i = \ln \sum_{j=1}^{M} e^{U_{ij}} + 0.5772 \tag{2.2}$$

where U_{ij} (also see Eq. 2.4) is the utility that angler agent i derives from recreational fishing at a recreational angling site j out of M sites. The variables and coefficient estimates in the *trip demand model* are presented in Raguragavan et al. (2013).

The *trip timing* decision is a discrete choice problem, with the choices being the days in the year when an angler starts their fishing trip(s). We used the timing information in the survey data to estimate a logit model for trip timing and this model is used in the agent-based simulation to determine the dates for fishing trips by a given angler agent i. The probability p_{ir} that the angler agent i starts a trip on day r among all possible sets of days s is given by the following logit formula:

$$p_{ir} = \frac{e^{\left(\sum_k \omega_k \cdot D_{kr} + \sum_l \sum_m \omega_{lm} \cdot X_{li} \cdot D_{mr}\right)}}{\sum_s e^{\left(\sum_k \omega_k \cdot D_{ks} + \sum_l \sum_m \omega_{lm} \cdot X_{li} \cdot D_{ms}\right)}} \tag{2.3}$$

where D_{kr} (or D_{mr}) is the k-th (or m-th) characteristics of day r, X_{li} is the l-th characteristics of angler agent i, and ω_k and ω_{lm} are coefficients to be estimated.

Trip length on the other hand is a continuous variable that takes a value of 1 or higher. We estimated a limited dependent variable model, Tobit, to provide a means of predicting fishing holiday lengths. Trip length in days (*TL*) is assumed to be a function of personal characteristics and the characteristics of the period during which the trip is taken:

$$TL = \sum_{u} \varphi_{U} \cdot D_{ug} + \sum_{v} \sum_{w} \varphi_{vw} \cdot X_{vi} \cdot D_{wg} \tag{2.4}$$

where D_{ug} (or D_{wg}) is the *u*-th (or *w*-th) characteristics of the trip start day *g*, X_{vi} is the *v*-th characteristics of angler agent *i*, and φ_u and φ_{vw} are the coefficients to be estimated. These trip length and trim timing model specifications are based on (Hailu and Gao 2012).

A random utility model (RUM) is used to predict angler preferences among a set of alternative sites. Fishing *site choice* is driven by cost of visit to the site, expected catch rates, the isolation score of the site, as well as other recreational attributes of the site. The most common RUM formulation is the multinomial logit (McFadden 1974), which provides the following closed form for the expression of the probability ($prob_{ij}$) that a person *i* chooses site *j* from *M* sites depending on the utilities expected from each of those sites.

$$prob_{ij} = \frac{e^{U_{ij}}}{\sum_{k=1}^{M} e^{U_{ik}}} \tag{2.5}$$

where, U_{ij} is the utility that angler *i* derives from fishing at site *j* and is dependent on site and angler characteristics as shown in Eq. (2.6).

$$U_{ij} = \alpha_0 + \alpha_1 \cdot cost_{ij} + \sum_{f} \alpha_f \cdot ECR_{ijf} + \sum_{k} \alpha_k \cdot S_{kj} \tag{2.6}$$

where α_0 is the base utility of a site, $cost_{ij}$ is the cost to angler agent *i* of recreational fishing at site *j*, ECR_{ijf} represents the number of fish of type *f* that the individual expects to catch at the site, S_{kj} stands for other site attributes that affect site choice (e.g. coastal length). Note that α_0, α_1, α_f, and α_k are regression coefficients. The estimation results for the model used here are presented in Table 9 in Hailu et al. (2011). The expected catch rates in the model depend on site attributes (particularly fish stocks) and the angler's experience. These rates are generated by another econometric model, the catch rate model, shown below in Eq. (2.7).

$$\ln ECR_{ijf} = \gamma_0 + \gamma_f \cdot stock_{jf} + \sum_{j} \gamma_j \cdot S_j + \sum_{i} \gamma_i \cdot X_i \tag{2.7}$$

where: ECR_{ijf} is the expected catch per trip of angler agent i at site j for fish type f; $stock_{jf}$ is the stock at site j of fish type f; S_j is the vector of other site attributes (such as if the site is man-made, if it is a beach, and so on); and X_i represents the demographic characteristics (such as age, education, employment, experience, whether the fish was a target species or not etc.) of angler i that influence expected catch. γ_0, γ_f, γ_j, and γ_i are regression coefficients obtained through econometric estimation. The catch rate functions used in our study are based on those reported in Table A5 in Raguragavan et al. (2013). We refer readers to (Gao and Hailu 2011b; Hailu and Gao 2012; Hailu et al. 2011; Raguragavan et al. 2013) for detailed model specifications and discussion of estimates.

The coral reef ecosystem model uses a local-scale model of trophic dynamics (Fung 2009) to describe interactions among algae, corals, and fish at a site. This model was originally developed as ordinary differential equations (ODEs) which have been parameterized as ranges in the Indo-Pacific region and the Western Atlantic region. Equilibrium behaviour and parameter sensitivity of the model have been examined in detail (Fung 2009). Since the coral reef ecosystem targeted (Ningaloo reef) has insignificant amounts of turf algae and sea urchins, this model has been simplified with only five functional groups, namely, hard corals (C), macroalgae (A), grazed epilithic algal community or EAC ($E=1-C-A$), herbivorous fish (H), and piscivorous fish (P). All the parameters in the coral reef ecosystem model are calibrated against recent observations of five functional groups in Ningaloo using a comprehensive learning particle swarm optimizer (Gao and Hailu 2010a). Details of the coral reef ecosystem can be seen in Fung (2009) and Hailu et al. (2011).

2.3 Overview: Framework-Specific Sequence

The ABM presented in this chapter has been applied to the assessment of alternative management strategies for recreational fishing in the Ningaloo coral reef marine park of Western Australia. Based on data from Tourism Research Australia on site surveys, it can be calculated that the number of tourists to the Ningaloo Coast for 2005 was about 203,580 (Schianetz et al. 2009). A recent survey report (Jones et al. 2011) in this area shows that 49 % visitors fish from the shore while 40 % fish from boat. This means that the model would work with a large population and it becomes necessary to develop a representative sample of the population being simulated.

Expert knowledge (EK) and participant observation (PO) are used to understand agents and their actions (M1). The expert knowledge used consists of the economic principles of utility maximization that govern angler choice among alternatives as well as scientific knowledge describing the dynamics of a coral reef ecosystem. Angler activities include choice of fishing site, choice of target fish, and expenditure on fishing related items such as bait. Detailed agent attribute data (M2) are elicited by conducting sample surveys, while agent behaviour (M3) derive from choice models, which are econometrically estimated based on collected survey responses. Further, we assume that the sample (collected responses of distributed surveys) used to generate attributes and behavioural parameters is representative, proportional up-scaling can be carried out, in which random sampling is used to generate the whole population (M5).

2.4 Technical Details

2.4.1 Data Summary

The work first conducted a survey of people who were fishing and recreating in the Ningaloo region of Western Australia, which is the target area of the study. The questionnaire was revised on at least two occasions. These revisions were based on feedback from staff members who visited the region and interacted with respondents who were willing to participate in the survey.

The survey questionnaire consisted of three sections, with the first two of these sections being the ones relevant for this study[1]. The first set of questions relate to the demographic details of the respondent and included information on country of origin, length of stay in the region, by what means the respondent travelled in the region as well as the size of the cohort with which the respondent was travelling. This section of the survey also collected information pertaining to the previous 12 month recreational and fishing experience in the region. For those who were fishing, information on the skill and experience of the angler as well as cost of the angler's fishing equipment was collected.

The second section of the survey asked participants to keep a log book of the fishing trips that they undertook to fishing sites in the Ningaloo region. The data requested in this section included: the site; the time at which fishing occurred; and the location at which the respondent lodged the night prior to the day of the fishing trip. The participants, when choosing a site, were asked to allocate a rank to a set of choice reasons and site attributes, i.e. they were asked to rank site scenery, importance of time availability, and other factors that might have affected their choice. Other information solicited through the survey included the species and number of fish caught and released and cost incurred as part of the trip (including the cost of bait, tackle, boat hire, boat fuel and food). Anglers were also asked to identify any fish species that they were targeting. It should be noted here that, as the data collection progressed, the log book approach was found to lead to low response rates. This is because the surveys were long and respondents had little incentive to fill out detailed information for multiple trips. Therefore, at a later stage in the data collection, it was deemed necessary that a face-to-face interview be used to improve response rates. The face-to-face interviews were conducted using the same questionnaire but

[1] The third section of the survey collected information on non-fishing recreational trips and was used for a separate study on non-fishing recreation.

it meant that one data point was obtained from each respondent instead of multiple data points as was initially hoped. The switch was successful and the project was able to generate a good enough sample through the face-to-face interview.

A total of 426 visitors were surveyed, and 402 of these provided trip information. Data collected covered a total of 774 trips. The data collected in the survey are stored in an Excel spreadsheet with a worksheet for each of the survey sections, i.e. demographic, fishing trips and recreational trips. Each section of the survey has been analysed and the results are reported in a summary report produced by Durkin (2009). These data described in Durkin (2009, p. 916) underpin the development of revised econometric models for recreational site choice and econometric models for fishing in Ningaloo.

During the initial data analysis, it became apparent that the survey information would be better stored in a database. The database not only records the information collected in the survey but also tables pertaining to fish species, the geographic distance between sites and a reference point, as well as information on respondents who were visiting in groups. This latter piece of information would enable a more accurate analysis of cost data for people fishing in groups. We refer readers to (Durkin 2009; Hailu et al. 2011) for further details on the data and analysis done on it.

2.4.2 Key Steps for Characterising and Parameterising Recreational Angler Agents

A recreational angler agent has demographic attributes (such as age, income, education level, employed status, and so on) and behaviour (such as choosing sites and catching fish). A fishing site is regarded as an agent that has environmental attributes (such as coral cover, algal cover, herbivorous fish biomass, piscivorous fish biomass, area, and coastal length) and ecological activities (interactions among dynamic environmental attributes). But in this chapter, we focus on the characterisation and parameterisation of recreational angler agents. So in this chapter, unless otherwise specified, “agents” refers to anglers. The structure of the econometric models that were estimated as a basis for empirically based behavioural models for recreational fishing anglers has been described in the “sub models” section above. Below, we provide an overview of the approach used.

The key steps involved in the econometric modelling of recreational choice and associated benefit calculation are outlined in Table 2.1 above. For recreational fishing, the first step is to obtain data on visitors and the choices they make. In our case, these data have come primarily from the survey conducted in Ningaloo. Data from the National Survey on Recreational Fishing was also used for a state-wide fishing study that included three sites in the Ningaloo region (Gao and Hailu 2011b; Raguragavan et al. 2012). In the second step, a theoretical model is used to provide a framework for describing observed behaviour or choices made. The key theoretical framework is the random utility modelling (RUM) framework for describing site choice. Other supporting models are specified using economic/econometric theory

Table 2.1 Research steps in econometric modelling and welfare change analysis

Research steps	Recreational fishing studies
Observe choices and profiles	National Survey of Recreational Fishing data (2000/2001) and Ningaloo fishing survey data collected by the project since 2007
Use a theoretical framework/model (RUM) and other empirical models	Five models: expected catch rate model, Site choice model (RUM), trip timing (logit model), trip length (Tobit model), and trip demand model (Negative binomial model). These models are grounded on economic/econometric theory and previous empirical evidence highlighting influences on choice
Estimate model parameters (econometrics/MLE)	Data fitted to models using maximum likelihood estimation (MLE)
Use model to predict behaviour and derive welfare values	Value of fish (part worth), value of change in fish stocks, site attributes, total fishing site values

and information from previous empirical studies. In the third step, econometric estimation is undertaken to generate the parameters of the RUM model and the econometric models listed in Table 2.1. Finally, the estimated models are used to drive agent behaviour and to calculate economic welfare change estimates arising from site condition or site management changes.

2.5 Lessons/Experiences

The major challenge in the research was that the rate of responses obtained for our survey questionnaires handed out in Ningaloo was initially low. This is because the data required was detailed and the survey was too long for respondents. As a result, the empirical analysis was delayed. Consequently, we changed our approach to data collection and began employing face-to-face interviews to maximize completion rates of the questionnaires. This change in approach enabled us to generate a usable sample that was bigger than was initially planned. What is more, the data from face-to face surveys were standardised so that names of fishing and recreational sites as well as sites of accommodation were checked for consistency in spelling. Where possible the location of the site is entered before the site name itself for easy identification, e.g. Exmouth Ningaloo Lodge, and this helped avoid naming confusions that would have occurred if we had relied on questionnaires filled out by anglers.

We were also aware that, before the data are used to parameterize our agent-based models, it was necessary and useful to analyse the survey data. Each section of the survey has been analysed using SPSS and the results are reported in a summary report produced by Durkin (2009). It should be noted that this analysis was carried out so that there is a better understanding of the strengths and weaknesses of the collected data. These data underpin the development of revised econometric models for recreational site choice and econometric models for fishing.

The parameterized models can be used to undertake evaluation of different management strategies such as the following:

1. Analysis of site closure effects
   The economic welfare losses from site closure can be estimated in detail, per person per trip. These values are based on estimated site access values. This method can also be used to evaluate the value of new fishing sites, as, for example, when anglers are provided with the opportunity to fish at a new site (e.g. made accessible through the construction of a road or a change in regulation). For the Ningaloo recreational fishing studies described above, estimated site values are shown in Hailu et al. (2011). For the state-wide recreational fishing model, these results are reported in Raguragavan et al. (2013).
2. Analysis of changes in site attributes
   One can look at increases or decreases in desirable site attributes. These calculations have been undertaken for fishing recreation in Raguragavan et al. (2013) using the state-wide fishing site choice model and for Ningaloo recreational sites in Hailu et al. (2011). The models presented above can be used to simulate welfare changes for different combinations of changes in site attribute values such as fish stock levels.
3. Integrated modelling of economic and biophysical effects
   The integrated ABM can be used to evaluate changes in outcomes in ways that take into account the feedback effects between economic choices (fishing) and fishing site conditions. Several demonstrative simulations of changes in management strategies have been undertaken and the results are published. For example, site access and fishing bag limit changes are simulated in (Gao et al. 2010; Gao and Hailu 2010b) while seasonal site closure regimes are simulated in (Gao and Hailu 2011b). In particular, the results reported in Gao and Hailu (2010b) indicate that it is possible for some restrictive access policies to be welfare improving even for anglers, because the stock gains and improved catch effects can outweigh the losses from reduction in access times. For further details, see (Gao et al. 2010; Gao and Hailu 2010b, 2011b).

Finally, ABMs and econometrically estimated choice models are popular approaches in the study of recreational behaviour. The two focus on similar data and are based on individual decision-making to determine patterns of recreational use. ABMs often rely on expert experience for defining rules to drive agent behaviours, while choice models govern individual behaviours using statistically estimated parameters. In our case, we regard choice models as a complement to ABM. Theoretical justification for agent behaviour structures in ABMs is weak in many contexts. The approached used in this study addresses this shortcoming by using theory and empirical data to define agent behaviour.

References

Durkin J (2009) Survey of fishing and recreation in the Ningaloo region. Unpublished technical report, School of Agricultural and Resource Economics, University of Western Australia, Perth, Australia.

Fung TC (2009) Local scale models of coral reef ecosystems for scenario testing and decision support. University College London, London, United Kingdom

Gao L, Durkin J, Hailu A (2010) An agent-based model for recreational management strategy evaluation: a case of Ningaloo marine park. The second international conference on Agents and Artificial Intelligence, Valencia, Spain

Gao L, Hailu A (2010a) Comprehensive learning particle swarm optimizer for constrained mixed-variable optimization problems. Int J Comput Int Sys 3:832–842

Gao L, Hailu A (2010b) Integrating recreational fishing behaviour within a reef ecosystem as a platform for evaluating management strategies. The 24th IEEE international conference on advanced information networking and applications, Perth, Western Australia

Gao L, Hailu A (2011a) An agent-based integrated model of recreational fishing and coral reef ecosystem dynamics for site closure strategy analysis. In: Chan F, Marinova D, Anderssen RS (eds) MODSIM2011, 19th international congress on modelling and simulation. Modelling and Simulation Society of Australia and New Zealand, Perth, Western Australia, pp. 3105–3111

Gao L, Hailu A (2011b) Evaluating the effects of area closure for recreational fishing in a coral reef ecosystem: the benefits of an integrated economic and biophysical modelling. Ecol Econ 70:1735–1745

Gao L, Hailu A (2012) Ranking management strategies with complex outcomes: an AHP-fuzzy evaluation of recreational fishing using an integrated agent-based model of a coral reef ecosystem. Environ Modell Softw 31:3–18

Gao L, Hailu A (2013) Identifying preferred management options: an integrated agent-based recreational fishing simulation model with an AHP-TOPSIS evaluation method. Ecol Model 249:75–83

Gao L, Durkin J, Hailu A (2010) An agent-based model for recreational management strategy evaluation: a case of Ningaloo marine park. The second international conference on agents and artificial intelligence, Valencia, Spain

Grimm V, Berger U, DeAngelis DL, Polhill JG, Giske J, Railsback SF (2010) The ODD protocol: a review and first update. Ecol Modell 221:2760–2768

Hailu A, Gao L (2012) Recreational trip timing and duration prediction. Tour Econ 18:243–251

Hailu A, Gao L, Durkin J, Burton M (2011) Ningaloo collaboration cluster: estimation and integration of socioeconomic values of human use of Ningaloo. Ningaloo collaboration cluster final report no. 4. http://ningaloo.org.au/www/dms/NingalooResearchProgramDocs/Publications/Cluster-final-reports/CP4_Socioeconomic_14June2011.pdf. Accessed 28 July 2013

Jones T, Wood D, Hughes M, Deery M, Fredline L, Jones R, Fulton B, Pham T, Pambudi D, Dwyer L, Spurr R, Chapman K, Lewis A, Chandler P, Catlin J (2011) Ningaloo collaboration cluster: socio-economics of tourism. Ningaloo collaboration cluster final report no. 3

McFadden D (1974) The measurement of urban travel demand. J Publ Econ 3:303–328

Raguragavan J, Hailu A, Burton M (2013) Economic valuation of recreational fishing in Western Australia: statewide random utility modelling of fishing site choice behaviour. Aust J Agri Res Econ. doi: 10.1111/1467-8489.12009 (AgEcon Search: http://ageconsearch.umn.edu/handle/97473. Accessed 27 July 2013)

Schianetz K, Jones T, Kavanagh L, Walker PA, Lockington D, Wood D (2009) The practicalities of a learning tourism destination: a case study of the Ningaloo Coast. Int J Tour Res 11:567–581

Schuhmann P, Schwabe KA (2004) An analysis of congestion measures and heterogeneous angler preferences in a random utility model of recreational fishing. Environ Resour Econ 27:429–450

Chapter 3
An Agent-Based Model of Tourist Movements in New Zealand

C. Doscher, K. Moore, C. Smallman, J. Wilson and D. Simmons

3.1 Model Description

As New Zealand's fourth largest industry, tourism plays a crucial role in the country's economy. Aside from providing jobs for thousands of New Zealanders, the foreign currency brought in by international tourists provides significant foreign exchange benefits and supports large portions of the rural economy. In the year ending in March 2010, tourism contributed NZ$ 6.5 billion to the economy, or roughly 3.8% to the country's total GDP (Ministry of Economic Development 2011). An important characteristic of tourism is its spatial and temporal nature. International tourists may spend anywhere between two days and several months touring the country, and in so doing, spread their economic, environmental and social impacts, both positive and negative, over a range of spatial extents, an impact we refer to as their "spatial yield". To gain a better sense of the effects of spatial yield as well as to provide a model to aid decision makers to optimise the benefits from tourism, we have developed an agent-based model of tourism movements around New Zealand which is grounded in decision-making data collected from tourists via semi-structured interviews. The primary purpose of the model was to give decision makers an opportunity to firstly understand and visualise the dynamics of tourist movements

C. Doscher (✉) · K. Moore · J. Wilson · D. Simmons
Faculty of Environment, Society and Design, Lincoln University, Lincoln, New Zealand
e-mail: crile.doscher@lincoln.ac.nz

C. Smallman
School of Business, University of Western Sydney, Sydney, Australia
e-mail: C.Smallman@uws.edu.au

K. Moore
e-mail: Kevin.Moore@lincoln.ac.nz

J. Wilson
e-mail: Jude.Wilson@lincoln.ac.nz

D. Simmons
e-mail: David.Simmons@lincoln.ac.nz

A. Smajgl, O. Barreteau (eds.), *Empirical Agent-Based Modelling – Challenges and Solutions*, DOI 10.1007/978-1-4614-6134-0_3,

and then to simulate the different impacts of a range of scenarios on tourists' spatial yield. These scenarios might relate to changes in the mix of tourists arriving in New Zealand, or fluctuations in currency exchange rates or fuel costs, as well as the influence of changes to the transport networks as a result of natural disasters or road closures. The model was designed to run on RePast Symphony 2.0 (North et al. 2006; ROAD 2009).

3.1.1 State Variables and Scales

International tourist groups (rather than individual tourists) are the primary agents in the model, interacting with each other and the environment. The groups (each referred to as a TravelGroup) varied in size and composition depending on the scenario being modelled and represent the major data input to simulations. After their arrival at one of two international airports (Auckland and Christchurch), tourist groups make a series of decisions about which activities to do and destinations to visit to compose their trip. Simulated events of meals, stays in accommodation, and movement within New Zealand are carried out as interactions with the environment, i.e., hoteliers are not explicitly present as independent agents, but the interaction is represented through methods which deduct an amount from a simulated tourist group's budget while increasing the amount in a simulated hotel's account based on a standard return rate. The design of the TravelGroups aims to represent the variety and temporal distributions of tourists visiting the country and required that generic TravelGroups be further classified by demographic values. Identifying those subclasses proved to be a significant challenge as there is a multitude of logical sets that could have been used. We were ultimately guided by a hierarchical (or cascading) approach to the outcomes of interview data to derive those sub classes. Since much of the analysis of tourism in New Zealand is broken down by nationality, the initial level of grouping attaches a nationality to each TravelGroup. At a later stage, this will also allow us to attach rough indications of available budget and daily spending habits (derived from an annual survey of international visitors, the IVS). In addition, nationality will allow us to predict the impact of culture on behaviour (Hofstede 2001). The initial version of the model simulated the period of 1 November 2008 to 28 February 2009 with inputs based on data from the IVS. This also allowed us to validate model outputs against real data. The type of trip groups were undertaking also proved to be an important variable identified from interviews, such that agents were initially grouped into Holiday, VFR Couple (Visiting Friends or Relatives) or VFR Family, Working Holiday or Round The World (RTW) classes with the distribution based on IVS data for the simulated period. The recorded length of stay "book ends" the trip and enables us to group the agents by the type of itinerary, being one of loop, triangle or stationary. Finally, attributes of each travel group are created for transport type and preferred level of accommodation, number of people in travel group, etc.

The following tables outline the agent and environment properties.

Agents: TravelGroups

Variable name	Brief description
Nationality	Text: records country of origin
Trip type	Text: records the purpose of the trip; holiday, VFR couple, VFR family, working holiday, RTW
Itinerary type	Text: type of travel pattern; loop, triangle, stationary
Transport type	Text: records the type of vehicle used; Petrol rental car, Diesel rental car, Petrol rental van, Diesel rental van, Public transport
Accommodation type	Text: records preferred level of accommodation; hotel, motel, B&B, backpackers, campground
Number of people	Number: records the number of people in the group
Trip duration	Number: total number of days for trip
Children present	Binary: 1 = children present, 0 = not
Budget	Number: records the total budget of the group
Itinerary	List: a sequential list of destinations
Kilometres travelled	Number: running total of kilometres travelled
CO_2 produced	Number: running total of CO_2 produced

For the baseline simulation, TravelGroups were sub-classed in a hierarchical fashion based on the distributions found in the IVS.

3.1.1.1 Environment

The environment encompasses the natural environment of New Zealand, including the road network and interconnected nodes (representing tourist destinations) along that network. NZWorld is the container for all environmental variables and also serves as a timekeeper for the simulations. Nodes are roughly equivalent to towns and cities, though not in all cases. For example, The Franz Josef and Fox glaciers are popular tourist destinations on the West Coast of the South Island. The town of Franz Josef is a base for many exploring that part of the country and is represented as a node. The Fox glacier is also a node as it is a popular destination but does not have any accommodation or restaurants and so is often visited by tourists on their way past the site. Nodes contain accommodation options and food outlets, though these are not explicitly modelled. Rather, each node has a set number of rooms available at each accommodation level, with node-specific average room rates derived from field work. No explicit food outlets are represented, but average prices for meals were derived from field work and those amounts are deducted from each TravelGroups budget at set times during the day based on which meal and the number of people in the group. Each node also contains a list of activities available at the node and includes cost, type of activity, and time required.

Environment: NZWorld

Variables	Brief description
Timestep	Number: time step
Time	Number: local time, derived from time step
Day	Number: day counter derived from time step
Node list	List: maintains the nodes
Road network	List: maintains the road network

Environment: Nodes

For each node a set level of accommodation types are available, distinguished by their cost per room. The cost as well as the number of rooms available at each level can vary from node to node based on field work.

Variables	Brief description
Name	Text: name of the node
Number of rooms (Accommodation type)	Number: number of rooms available for each level of accommodation (hotel, motel, B&B, backpackers, campground)
Accommodation rate (Accommodation type)	Number: cost per room for each level of accommodation
Meal cost (Meal type)	Number: cost per meal for each level of meal (breakfast, lunch, dinner)
Activities list	List: manages activities available
Stayers list	List: manages TravelGroups staying at that node
Travellers list	List: manages TravelGroups travelling to their next destination
Departers list	List: manages the TravelGroups departing the country

Environment: Nodes: Activities

Each node carries a list of activities that tourists can choose from, which could include anything from nature walks, to bungy jumping to more seasonal activities like skiing. Input data were collected during field work. Some activities can be weather dependent and a flag indicates if this is the case. In future versions, this flag, coupled with daily weather conditions will be used as a decision parameter by TravelGroups.

Variable	Brief description
Name	Text: name of activity
Cost	Number: cost per person for activity
Time	Number: hours required to do activity
Weather dependent	Binary: 1 = yes, 0 = no

Environment: Road network
TravelGroups travel from node to node along the travel network

Variable	Description
Road ID	Number: unique ID for each road segment
Road length	Number: length of each road segment (m)

3.1.2 Process Overview and Scheduling

The model uses an hourly timestep with each day roughly composed of 15 timesteps. The NZWorld object serves as the environment, timekeeper and messenger. At set points during the day, lists of TravelGroups at each node are messaged to have the appropriate meal for that time and either choose their next activity or travel to their next destination. Each day begins at 9.00 AM. All TravelGroups in the travelling list "have breakfast" (an amount is deducted from their budget based on the cost of breakfast at that node and the number of people in the group) and check out of their accommodation (an amount is deducted from their budget based on the number of nights stayed, number of rooms used and room cost). Groups travelling that day then determine the best route to their next destination and travel there. Agent movement is based on Nick Malleson's RepastCity model (Malleson 2011). Upon arrival, groups "check in" to a preferred level of accommodation by querying the node's list of available accommodation. If no rooms are available at that level, the next level down is queried until rooms are found. Groups then check the local time and wait for a message to have a meal and chose an activity. Groups in a node's staying list choose their next activity. Each day ends at 11.00 PM local time. At this point, NZ-World updates the travelling and staying lists by comparing the current destination to the next day's destination—if they are different, that group is transferred from the staying list to the travelling list. The next day begins at the next time step. TravelGroups may also be identified as departing and are shifted to the departing list.

Upon departure, each group is queried for its itinerary, budget and CO_2 produced which are the used to establish spatial yields. These data can then be compared against the IVS data in the baseline scenario's case, to evaluate how well the model has performed. For other scenarios, this report allows us to monitor their activities and destinations. At the end of the simulation, nodes are queried to estimate economic yields based on the number of groups that stayed there over the simulated periods.

3.1.3 Design Concepts

The model was designed to simulate the spatial yields of tourist movements to the New Zealand economy and environment. At this point, the agents are not adaptive in the decision making process.

Collectives An important design consideration has been simulating groups of tourists rather than individuals, a decision that was supported by survey data, where it was observed that decisions were made in a group context rather than being driven by an individual (unless one of the group members was a New Zealander or someone with previous travel experience in New Zealand). The survey data also pointed out the importance of group composition on decision making, particularly the presence of young children.

3.1.4 Details

On startup, the model builds up a simulated New Zealand environment as an NZWorld object composed of the road network and nodes. The locations of nodes are read into the model as an ESRI point shapefile and additional data (accommodation, activities, restaurants and accompanying costs) are read in from a comma separated (CSV) file at the beginning of each day. Travel group arrivals and their attributes are also read in to the model from a CSV file. NZWorld also functions as a timekeeper of the simulation, breaking the simulated day up into three primary time blocks (morning, afternoon and evening) for meals and activities. Once instantiated, agents check into a virtual hotel for their first day and are endowed with a total budget, a daily spending limit, a preferred level of accommodation and a list to hold activities already done. During the development phases, the group's itinerary is set based on the total number of days in their trip, their arrival and departure nodes and their itinerary type (e.g., loop, triangle, or stationary).

The input data constituted a significant portion of the modelling effort. For agent instantiation it was necessary to derive an input data set that was representative of tourist arrivals for the simulated baseline period. The data set would need to provide the sub class parameters (nationality, trip type, itinerary type, etc.) on a daily basis and split by the two main international gateways (Auckland and Christchurch). These data were derived from the IVS datasets.

There are essentially two submodels at work in the model; one to simulate the in-node behaviour of staying TravelGroups and the other for between-node behaviours. Upon arrival at a destination, groups switch from the between-node to the in-node models.

3.2 Overview: Framework Specific Sequence

Our intent with this model was that it would be grounded in the context of actual decisions made by tourists. Parameterisation methods from steps M1, M2 and M3 were used in model development. These methods are described in more detail below.

3.2.1 M1 Model Characterisation Methods

In this version of the model, only TravelGroup agents where explicitly modelled so the agent classes were primarily determined using expert knowledge and a review of existing literature.

3.2.2 M2 Attribute Data Elicitation Methods

TravelGroup attributes were based on the pre-existing International Visitor Survey as well as being informed by a previous study that focused on TravelGroup characteristics for a particular area of New Zealand (the West Coast of the South Island, Moore et al. 2001).

3.2.3 M3 Behavioural Data Elicitation Method

The behavioural aspects of TravelGroups were primarily influenced by semi-structured interviews carried out *in-situ* at several tourist destinations.

3.3 Technical Details

Rather than using these sets of methods sequentially, we found it advantageous to develop the agents and their behaviours in an iterative fashion, employing methods when it was most obvious to apply them. Below, more detail is provided on how each method was used.

3.3.1 M1 Methods

3.3.1.1 Expert Knowledge

In addition to the expert knowledge engendered in the research team, a series of meetings were held with representatives from key stakeholders in the New Zealand tourism industry including the New Zealand Tourism Council, the Department of Conservation, and the Ministry of Tourism. The key requirements of model outputs were identified and the model of tourism behaviour was iteratively developed incorporating the perspective of those supplying the tourism experience to visitors. In essence, these insights were the result of observations of tourist behaviours from their point of view with a very large sample size. These discussions also influence the structure of the model. For example, structuring the model into three main blocks of time during the day was suggested by a representative of the Tourism Council based on their experience. The experts also provided suggestions on how the agents could be sub-classed to achieve a more realistic representation, allowing us to ensure that model outputs were useful and meaningful to the users. In addition to industry representatives, we also conducted meetings with "front line" tourism staff at tourist information centres. These meetings also provided valuable insights into tourism behaviour, such as the observation that most tourist itineraries appear to be set in advance and are not often modified after arrival in-country.

3.3.1.2 Literature Review

Existing literature was used to develop behavioural aspects of the agents. Previous studies from New Zealand's West Coast (Moore et al. 2001) provided high resolution temporal data on how tourists made decisions about where to go and what to do.

3.3.2 M2 Methods

3.3.2.1 Survey

The International Visitor Survey (IVS) is carried out annually by the Ministry of Economic Development. Departing international visitors are surveyed on questions regarding their travel patterns and expenditures, including places visited activities or attractions undertaken, accommodation choices and transport modes broken down by nationality, country of origin, and tourist type. The value of this dataset was twofold: in the first instance, the IVS allowed us to formulate the breakdown of incoming tourists (including their lengths of stay and entry/exit points) and also provided a set of data to validate the simulated tourist behaviours against. This survey is undertaken and carried out annually independent of our research team. Survey results are available in the form of Excel spreadsheets with each tourist group have a set of records for each destination. Entry and Exit dates and locations are included along with number of nights at each destination, expenditures on food, activities, and accommodation, mode of transport. A limitation of these data is that while they provide the location of destinations, they do not include the route each group took to arrive at a given destination, which can influence between-node activities. (More detail can be found at www.met.govt.nz/sectors-industries/tourism/tourism-research-data/international-visitor-survey/.) While the survey results provided crucial parameterisation data, additional analysis was required to formulate arrivals of the TravelGroups and representative itineraries (an aspect also informed by data from "front line" staff). Restructured IVS spreadsheets then became inputs to the model which detailed the arrival and breakdown of TravelGroups on a day-to-day basis.

3.3.3 M3 Method

3.3.3.1 Semi-Structured Interviews

As the primary aim of this model was to be grounded in actual tourist behaviour, *in-situ* interviews were crucial to model development and parameterisation. As these interviews provided the bulk of the information used for deriving the conceptual model of agent behaviour, we will go into more detail on this aspect of the parameterisation. The interview data were contextually analysed both manually and electronically (using NVivo software) to look for common themes in decision making and important variables that influenced the process. Lengths of stay of respondents ranged from six days to one year (Moore et al. 2009). An important variable that emerged from the data analysis was that tourists could be categorised by their "Trip Type" related to the purpose of the visit. The primary categories of Type of Trip are sightseeing, and visiting friends or relatives (VFR), holiday/family, working holiday and 'round the world' (RTW). The category a tourist group was classified into tended to correlate well with the style of travel, itineraries, transport and accommodation choices. Another important result suggested that, in general, the first third of their trip was more planned while the middle and final thirds were less planned and structured, indicating that tourists' decision-making evolved as they became more familiar with the ease of travel in the country. In addition, during the initial third of the trip, accommodation choices were more or less "locked in" and became more open ended as the trip progressed.

As part of the model design process, 140 interviews were carried out by intercepting tourists at five separate locations in the Canterbury region of New Zealand's South Island. Analysis of these interviews provided insights into the factors that affect tourists' decision-making processes and allowed us to develop heuristics for modelling purposes.

The primary focus of these interviews was to identify the key drivers of the decision-making process by probing how they came to decisions for such choices as their destinations, their overnight accommodation and the activities they took part in. Demographics including gender, age, nationality, country of residence, travel group details, type of transport, and length of stay were also collected. The sites chosen for the interviews represented different destination sites: a 'gateway' or entrance point into New Zealand (Christchurch), a 'terminal' or diversion from a main road site (Akaora and Hanmer Springs), and a 'through-route' site located on a main through-route (Kaikoura and Tekapo). This allowed us to sample tourists at various points in their trip as well as at different destination types. Based on the experiences of the interviewers and a review of the results, a second interview protocol was developed which focused only on what tourists had done on the previous day or immediately before the interview.

The interview analysis also suggested that there were four important dimensions of decision-making that could influence tourist behaviours: (In)Flexibility; timing/location; social composition; and stage of trip (as just discussed). (In)Flexibility relates to the perceived 'ease' of travel in New Zealand, the level of decision

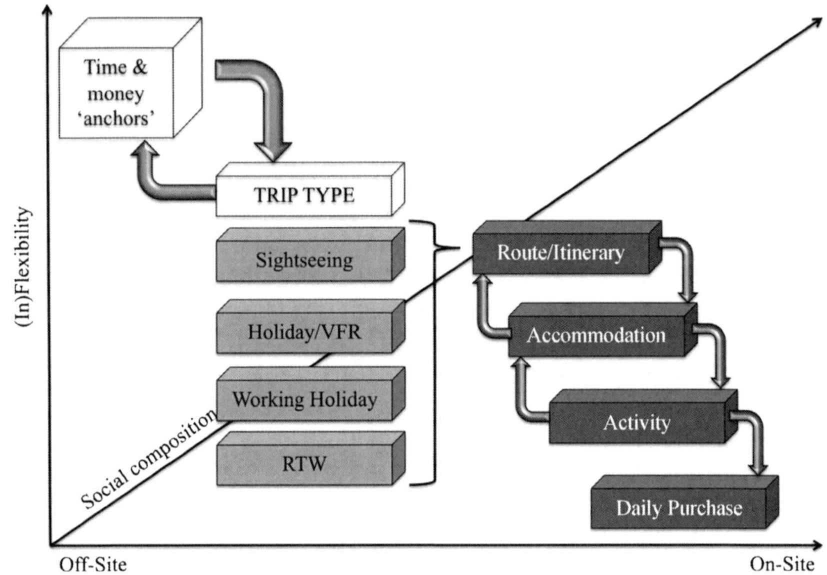

Figure 3.1 Three dimension cascade model of tourist decision making

openness and the openness of tourists to advice and information. The dimension of timing/location calibrates when and where decisions were made and includes the number of decisions that may have been made before arriving and could be influenced by the perceived risks of not booking accommodation or activities in advance as well as the needs of the particular group members. 'Social composition' concerns the range of social influences upon the decisions made by tourists or groups of tourists. In particular, it represents the effect that travel group composition has on decisions made, such as the presence of young children, or the presence of either a New Zealander or a previous visitor in the group. It also includes, however, the effect of others beyond the group (e.g., other tourists, locals encountered and friends and relatives in New Zealand but who may not be travelling with the tourist or travel group). Additionally, the observation that the decision making process appears to change with length of time in the country, reflected in the stage of trip, allows us to breakdown trips into equal thirds, typified by different styles of decision making.

This analysis allows us to propose a 'cascade' model of decision-making where the Type of Trip a tourist group is on leads to a particular cascade of decision-making that is influenced by the remaining three basic dimensions of identified above (see Fig. 3.1).

Available time and budget anchor or bookend the trip and, importantly, determine the Trip Type prior to arrival in New Zealand. Once decided, this provides a framework for the subsequent classification of agents in the model.

An example should help to illustrate how these dimensions interact to produce tourist decisions and behaviour. A travel group on a 'holiday/family' Type of Trip (e.g., often from Australia for a short duration) may well be relatively inflexible (i.e., high '*Inflexibility*') about accommodation and itinerary choices because those decisions were made prior to travel and off-site (i.e., early/off-site '*timing/location*') solely by the adults/parents (i.e., low/closed '*social composition*'); activity decisions may be partly inflexible (i.e., moderate '*(In)Flexibility*'—e.g., a child in the family wishes to do the 'luge' at Rotorua but some other activities remain flexible) and have relatively high '*social composition*' (e.g., other family members influence the decisions as may recommendations of locals or New Zealand friends/family); further, activity decisions may not be finalized until on-site (i.e., late/on-site '*timing/location*') but channelled by prior information (and the 'luge' preference of a family member); daily purchases, including what and where to eat may be relatively flexible (i.e., moderate to high '*(In)Flexibility*'—though constrained, perhaps, by cost and dietary preferences and needs) and open to influence by locals (e.g., the recommendation of the owner of the motel where the group is staying); even during a relatively short stay some flexibility may be built into the middle part of the trip (e.g., have a day trip to *either* destination A *or* B at about mid-trip).

These dimensions and their magnitude in relation to particular types of decisions (e.g., accommodation, activities, itinerary, etc. decisions) made, as they are, within particular Types of Trip then interact with the decision making process to affect the kinds of heuristics that are 'chosen' to make—or simply emerge as the means of resolving—particular moment-by-moment decisions. That is, the heuristics that have been identified in the extensive literature on human decision-making are effectively 'sieved' through the matrix created by the dimensions identified from the data in the qualitative interviews in relation to particular types of decisions within the context of particular Types of Trip. In the case of the 'holiday/family' Trip Type, for example, relatively deliberative heuristics (e.g., elimination by aspects) may be used to come to activity decisions that have moderate flexibility, involve relatively high social composition (e.g., other family members) and are likely to be made in New Zealand but prior to arrival at particular sites. By contrast, daily purchases that are highly flexible, low in social composition (e.g., up to an individual) and made at the point of the behaviour (e.g., in a shop) may be determined by less deliberative heuristics (e.g., take the best immediately on offer within a set price range). Inspection of the IVS revealed that typical itineraries could be placed into one of three groups: loop (multiple nodes starting and stopping at the same node), triangle (three nodes starting and stopping at the same node) or stationary (staying primarily at the same node but making short trips out and back).

An important point is that the nature of this dimensional matrix will be specific, in the model, to each kind of decision for each kind of Type of Trip. In addition, it will be affected by the particular 'third' of the trip within which that decision is being made. The fieldwork was also designed so that the qualitatively derived 'agent categories' of Trip Type could be 'cashed out' in 'fuzzy sets' of variables (such as nationality, length of stay, repeat visitation, age, transport type, accommodation type, etc.) that have been used as the basis for standard data collection on international tourists in New Zealand (principally, the International Visitors Survey (IVS)).

Informed by these insights, a cascade of decision making events was designed and implemented.

3.4 Lessons/Experiences

The parameterisation of the agents encompassed a significant portion of model development and required a multi-disciplinary team to best represent the system. A social scientist drove the survey stages and provided a theoretical framework for decision making, a tourism expert provided a crucial perspective based on experience, and the modeller translated this into workable code. Interviews with front line tourism staff allowed us to refine our model and "ground truth" its outputs. In addition, the views and perspectives of the tourists themselves provided the "grounded" data that drove the modelled decision making process. While the interviews occupied a significant time block of the project, the process of determining the best configuration of the agents posed significant challenges. As noted previously, there are innumerable ways in which the agents could be classified and many were considered. Our final decision was guided by the fact that most measures of tourism impact were tied to nationality and transportation type and so these were used as primary classifiers. Thus, our final classification was chosen to facilitate comparisons of the outputs with available data and studies rather than to provide an alternative or novel view of tourists. This doesn't necessarily mean that we are unable to review the results from a different perspective (i.e., from a different classification) and once the model is further developed, we may need to revisit these classifications to see if alternatives provide any additional insights into how visitors to New Zealand make their decisions and the impacts those decisions have on the economy and environment.

The interviews provided the most significant inputs to model development. Towards the end of the interview period, an interviewer began to ask questions along the line of "tell me what you did yesterday" in addition to the questions in the protocol. The responses were valuable as follow-up questions (e.g., "Why did you decide to do that?") provided insights into not only what decisions were made, but how they were made. In retrospect, we would include more questions like this, that allow tourists to describe their decision making process in their own words, which could allow us more information to make more general conclusions about how a specific type of tourist makes a decision.

As noted earlier, the research team spent a great deal of time discussing the sub-classing of agents. Our final classification was influenced by the desire to compare model outputs with existing datasets (the IVS in particular). On further consideration, it's apparent that agent behaviours are more important than those classifications, and we would encourage modellers to focus on the former over the latter (though acknowledging that different behaviours can be tied to different attributes/classifications), since the model outputs can be restructured to more amenable forms for particular analyses.

At this stage, the model is a simplistic but effective simulation of tourist movements. The visualisation of the TravelGroup movements reflects the pulses of groups that emanate from destinations on a daily basis and travel times approximate well to the observed times. There are several limitations to the model in its current state, the most significant of which lies in the current inability of the agents to independently choose their own destinations. Our survey data indicate that, for most visitors, the itinerary is set in advance as trip duration "bookends" the trip and limits the number of places that can be visited. For many visitors, particularly those that are in New Zealand for extended periods, destinations choices are made *in-situ* using information gained from other tourists or from locals. It would be beneficial to implement a higher level of interaction between other tourists which would allow more dynamic decisions to be made. It may well be that emergent patterns of destination choice then appear as "word of mouth" could influence when and where groups choose to go.

References

Hofstede G (2001) Culture's consequences, comparing values, behaviors, institutions, and organizations across nations. Sage Publications, Thousand Oaks CA

Malleson N (2011) RepastCity—a demo virtual city. http://portal.ncess.ac.uk/wiki/site/mass/repastcity.html. Accessed on 7 June 2010.

Ministry of Economic Development (2011) Key tourism statistics. http://www.tourismresearch.govt.nz/Documents/Key%20Statistics/KeyTourismStatistics2011(July).pdf. Accessed on 8 February 2010.

Moore K, Simmons DG, Fairweather JR (2001) Visitors to the West Coast: characteristics, attractions and decision-making. Moore Tourism Recreation Research and Education Centre (TRREC), Lincoln University, Report No. 22

Moore K, Smallman C, Wilson J, Simmons D (2009) Yield-relevant Tourist Decision Making. LEaP Report No. 11, Lincoln University, New Zealand

North M, Collier N, Vos J (2006) Experiences in creating three implementations of the repast agent modelling toolkit. ACM Trans Model Comput Simul 16(1):1–25

ROAD (Repast Organisation for Architecture and Design) (2009) Repast home page: http://repast.sourceforge.net. Accessed on 27 July 2010.

Chapter 4
Human-Ecosystem Interaction in Large Ensemble-Models

Randall Gray, Elizabeth A. Fulton and Richard Little

4.1 Introduction

Managing the effects of human activities on environmental assets must surely be one of the most vexing jobs imaginable. The nature of human interactions changes with social, economic, political and technological changes and management strategies are implicitly difficult to assess: as soon as one is in place, the baseline for comparison changes. Unfortunately with growing population and associated demands it is an issue that cannot be avoided.

Western Australia is experiencing a significant increase in the level of natural resource extraction and export (ABS 2012), and has long standing industries in fishing and agriculture. Tourism is steadily growing (ABS: cat. 8635.0.55.002) and there is a corresponding increase in the demands on public infrastructure. Many of the public uses are associated with the state's reefs and its marine environment, including protected areas such as the Ningaloo Marine Park which was listed as a World Heritage site in June 2011.

Finding ways to equitably manage the environmental assets that these industries are dependent on is a complex task. There are a great many possible kinds of uncertainty about the modelled system (Francis and Shotton 1997), and there are likely to be multiple uses, many of which may have conflicting needs or measures of amenity (Grumbine 1994).

The natural resources used by society are part of a larger system, and their use can have important consequences beyond those which are immediately concerned

R. Gray (✉) · E. A. Fulton · R. Little
The Commonwealth Scientific and Industrial Research Organisation (CSIRO), Hobart, Australia
e-mail: Randall.Gray@csiro.au

E. A. Fulton
e-mail: Beth.Fulton@csiro.au

R. Little
e-mail: Rich.Little@csiro.au

A. Smajgl, O. Barreteau (eds.), *Empirical Agent-Based Modelling – Challenges and Solutions*, DOI 10.1007/978-1-4614-6134-0_4,

with their use. Inadequate management strategies may have unintended consequences or cause damage that is difficult or expensive to remediate. Management strategy evaluation (MSE) frameworks use simulation models to test and compare the likely outcomes of management strategies to highlight, and help avoid, such outcomes (de la Mare 1998). This approach inherently deals with uncertainty and multiple objectives, and allows for alternative strategies to be compared against a common baseline.

Strategies under consideration may be based on a wide range of management levers and have multiple competing objectives. Strategies are often adaptive and typically include environmental assessment as well as socio-economic considerations. In turn these feed directly into determining what management decisions are made, and the implementation of the resulting regulatory actions including uncertainty about the effectiveness of implementation (e.g. non-compliance).

Robustness of the alternative strategies to system uncertainties is gauged by considering their performance across simulations with alternative sets of parameters and drivers. Information elicited from the simulations can also be incorporated into real world decisions, to improve strategies by recognising potentially significant drivers in the system or by making particular failings in the strategies more obvious.

While some MSEs are performed using bespoke models, more generic frameworks have also been developed. For example, a hybrid modelling platform InVitro (which brings together concepts from metapopulation, differential equation and agent based modelling) has been developed expressly for the implementation of integrated MSE which applies a management model to systems as a whole, rather than disjoint subsystems. The framework has been applied in two regions of the north west of Australia (Fig. 4.1). Although MSE has previously been applied to the management of different industry sectors individually (Butterworth and Punt 1999; Milner-Gulland et al. 2010), these studies applied MSE to the management of the ecosystem and to a suite of industry sectors.

The purpose of this chapter is to describe the models used to represent various aspects of human behaviour across a range of sectors, against the backdrop of a model of the biophysical components. Before discussing how to parameterise such large system-level models, it is important to understand the general design of the model so this chapter provides a high level ODD description of the model before using the framework from Chap. 1 (and Smajgl et al. 2011) to discuss the model parameterisation. Such a large model cannot be fully described in a single chapter and for interested readers further details can be found in the technical reports associated with the two studies (Gray et al. 2006; Fulton et al. 2006, 2011).

4.2 Overview

InVitro is an agent-based framework for constructing simulation models that integrate the dynamics of ecosystems and human activities. It has been used to implement two models of large marine ecosystems and associated industries along the northern and western coasts of Western Australia, specifically the Pilbara region (Gray et al. 2006) and the Gascoyne region (Fulton et al. 2011).

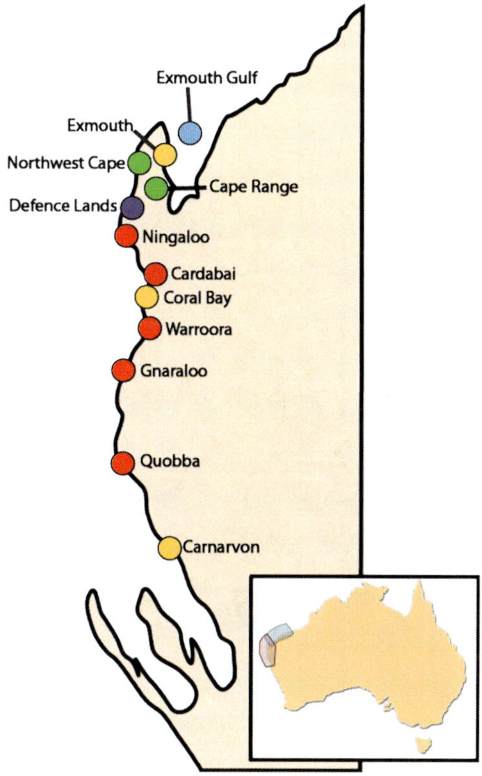

Figure 4.1 Schematic map of the tourism node network used in the Gascoyne model, with smaller insert map showing the general regions of Australia modelled (the *blue* region is the Pilbara and the *darker red* area the Gascoyne). *Orange* nodes in the larger Gascoyne map are settlements, *red* are pastoral stations, *green* are national parks, *blue* are entirely marine and *purple* are effectively unrestricted at present

A number of industries and activities put pressure on the environments modelled in these studies, including petroleum exploration and extraction, tourism, coastal development (and associated urban infrastructure, services and amenities), agriculture, shipping and road transport, salt production, port operations and both commercial and recreational fishing.

The Pilbara model took an "intermediate complexity" approach to representing system structure. It concentrated on a few species of scale-fish, prawns and sharks, and the biogenic habitat the species are dependent upon. The anthropogenic components focused on contaminant effects associated with salt production and gas extraction, as well as the physical effects of trawling, trapping, dredging, transport and the presence of pipelines and oil rigs.

The Gascoyne model was considerably more complex; both its ecosystem and the models of human activity are richer in detail (e.g. including climate and biogeochemical drivers, trophodynamics, agent based representation of the many industries and a model of the regional economy) and each of these agents has more scope for interaction amongst the other model components. The major industries represented in this system model are commercial and recreational fishing, tourism (which includes accommodation, charter boat activities, local spending and road use), petroleum

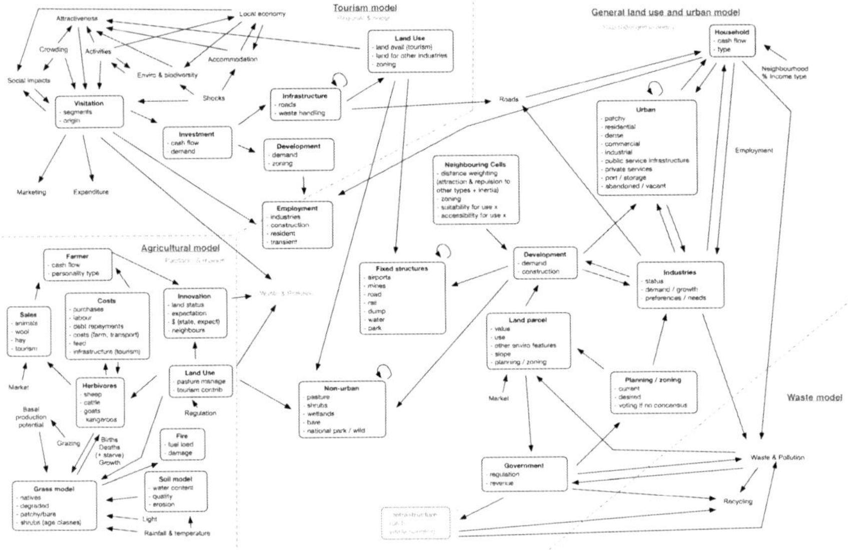

Figure 4.2 Schematic of the major components of the terrestrial human use models

exploration and production, terrestrial transport, agricultural production and an urban demographic, development and economic model (e.g. see Fig. 4.2).

4.2.1 *Entities*

The ecosystems and environments can be seen as both a stage for social, industrial and economic activity, and as the basis for the assessment of the efficacy and robustness of different strategies in the two studies. Many of the entities in the system can be represented by a number of alternative models with different levels of aggregation or "process" knowledge about the level of "life-history".

In order to function as a component in an InVitro ensemble, each agent (an instance of a model in the system) must maintain a small set of state variables in addition to those related to its internal functioning. This set includes a unique identifier, the agent's age, its "current" time, a default time-step, and a list (which may be empty) of times at which it must begin a time-step.

4.2.1.1 Representations of the Physical Environment

The components of the physical environment may be represented in a number of ways. For example they may be values associated with grid-cells or polygons which are read in from external data, or they may be attributes generated by equation based representations, which are conditional on other environmental state variables. The fundamental environmental features are geospatial, time-stamped, attribute layers which may remain constant throughout an entire simulation or change through time.

4.2.1.2 Ecosystem Representations

The ecosystem represented in the model includes the major species (or food web components) pertinent to the human activities of interest (including conservation, which is why the Gascoyne food web was so extensive). These species are selected based on abundance or biomass surveys of the system (i.e. dominant species that characterise the top 85–90% of the biomass in the system), network analysis of data on diets and habitat dependencies and expert ecological advice about system structure and key dynamics. Clustering methods like regular colouration (Johnson et al. 2003) help identify useful levels of ecological aggregation when creating the ecological structure of the model. In the Gascoyne region the close connection of coastal terrestrial and marine activities meant that it was important to extend the ecological representation to terrestrial habitat (pasture and bush), domestic and feral livestock and key native fauna (macropods).

Species may be represented in a simulation in several ways, to best capture the life-history of the organisms. Each representation has its own set of state variables and native scales appropriate to the processes modelled. For instance turtles are represented by difference equations for patches of eggs on nesting beaches (i.e. aggregate clutches for the whole beach), metapopulations of small pelagic juveniles, localised age structured populations of sub-adults and then small (agent-based) groups of adults (in some instances individual adults are followed). The essential features of all these representations however are age, a biomass and abundance, natural mortality, reproduction, trophic interactions and habitat affinities. Similar mix-and-match approaches can be used for all species, though typically a gridded model domain is used for the planktonic or benthic and terrestrial habitat components of the system, broad scale patch models are used for benthic invertebrates, metapopulation models are used for forage fish, local age structured population models for reef fish and terrestrial herbivores, schools for large pelagic predatory fish and sharks, and individual models for large bodied animals like whales, dugongs and whale-sharks.

4.2.1.3 Anthropic Models

While considerable effort was put into the appropriate representation of the biophysical system components, the aim of both studies was to support sustainable multiple use management. This meant that each of the major industries had to be represented in the model, particularly their regulation, production and environmental impacts. In both studies sets of management strategies that treated each sector independently were contrasted with strategies that required coordinated management across sectors. This meant that simulated analogues of detailed monitoring and regulatory models were required, along with departmental communication and political lobbying models (to capture the implications of political impediments to management integration).

One of the most detailed sectors in both InVitro models is fishing (commercial, recreational and charter). This is because it is a major source of human induced perturbation in the coastal environment (Blaber et al. 2000; Nellemann et al. 2008). While the fishing sectors in the Pilbara tend to operate in separate geographic locations and the main interactions are at the regulatory level, in the Gascoyne the sectors co-occur and cross-sectoral economic and social influences dictate the types, levels and distribution of fishing effort. As with the ecological components, human activity can also be represented by different models. Recreational fishing is a good example, as the representation of this effort is quite different in the two studies. For the Pilbara model a probability of catch is assigned based on the size of the human population in the residential centres, attenuated both by distance along road networks and with distance from locations at which boats can be launched. In contrast, the Gascoyne model explicitly simulates individual effort, or groups of individuals on charter boats, based on the number of humans in the region (local residents or tourists), desired catch characteristics and their expectation of being met at different locations (in turn based on information from direct experience by the agent or via information shared when agents meet).

Petroleum production and associated port facilities are also important features of both models, as they are strong drivers of population size and anthropic effects in the regions. In the Pilbara study the petroleum sector's role is limited to the effects of its physical infrastructure and transport, while in the Gascoyne, specifics of the production as well as the sector's economic and social roles are also included. The levels of petroleum production in both studies influence the amount of shipping traffic with concomitant effects on congestion around ports and infrastructure requirements. Ports are the start and endpoints for ship movement in both models, and are also associated with human populations and port related sectors.

The rest of the anthropic activities in the Pilbara study are represented by impact models, whose effect is only to alter conditions or population levels. These other sectors are more detailed in the Gascoyne model, including:

- A complete range-lands production and management model with attitude profile and friendship networks. The economic status of the pastoral stations dictates the activity mix (pastoralism:tourism) and the kinds of investments undertaken (maintenance and new infrastructure builds);
- A hybrid cellular automata/agent-based demographic and services model representing the resident human population centres. This model tracks current land-use, zoning, development and current amenities and infrastructure at a quarter-acre block level. The human population is modelled at a 1-to-1 level, tracking age, income, education type and level, employment type and status, location of home and work places, household type, family status;

- An agent-based tourism model that tracks the make-up, routes, accommodation and activities of tourists in the region as well as their expectations, satisfaction, information sharing and expenditure; tourism regulation is also modelled. This model operates at group sizes from individuals up to coach tours;
- A road transport model which includes use of accommodation by drivers, demand for mechanical services and road kill;
- A simple salt mining module (mainly an impacts model, but also with employment, housing and service demand components);
- A regional economic model based on an input-output framework that is dynamically populated by all of the industries operational in the appropriate region of the larger model.

4.2.2 Process Overview and Scheduling

InVitro can be viewed as being similar to a computer operating system which runs agents instead of programs. Each agent is scheduled throughout the simulation period to run at various times which are determined by its own subjective time, its default time-step and interactions with other agents. Time is treated as a continuous variable, with no constraints other than that it must increase monotonically as the model ensemble steps through time.

Most agents are rostered into the "run-queue" which is a priority queue sorted primarily on the subjective times of the agents. Each agent also has a priority which governs the sorting within a group of agents slated to begin at the same subjective time. This makes it possible to ensure that agents like cyclones have an effect on all the appropriate parts of the system. Optionally, a third sorting key may be included which randomises the order of dispatch in a set of agents with a common subjective time and priority.

Some agents must run synchronously with all the other agents in the ensemble, and these are rostered into the "standing-queue". Standing-queue agents are often present as a bounding or forcing function, or provide essential information. Examples include things like temperature, cloud cover, or rainfall. Algorithm 4.1 demonstrates the way these queues interact.

while *there is an agent in the run-queue with a subjective time,* st_{rq} *<end-of-run* **do**
    for each *agent in the standing-queue* **do**
        if $st_{sq} <= st_{rq}$ **then**
            run the agent with a time-step of $st_{rq} - st_{sq}$
        end
    end
    run the agent at the head of the run-queue
    if *save-snapshot* **then** save state of the whole model
    if *halt-raised* **then** halt the whole model
end

Algorithm 4.1 Main simulation loop. Subscript *rq* and *sq* indicate association with either the run-queue or the standing-queue respectively

Each agent sets its time-step to whatever positive value seems best: the duration is usually either some parameterised default value or some value which is based on the agent's state and the states of the agents which comprise its local environment. Variable time-stepping causes drift in the relative subjective times of the agents comprising the ensemble, but Algorithm 4.1 ensures a monotonic flow of time.

4.2.3 Design Concepts

The size of the geographic domains associated with major human activities in marine environments made the prospect of representing everything at an individual level unappealing. Nevertheless, both systems included aspects that were important to capture at an individual level (such as contaminant contact, or whale-spotting); so the level of aggregation of animals in the ecosystem needed to span the range from individuals to localised sub-populations. InVitro was designed with these constraints in mind, and it provides a suitable vehicle for these regional MSE studies.

For consistency, the decision-making and responses of the anthropic models needed to be flexible and dynamic. This was accomplished by allowing assessments, response selections and response intensities in these models to be configured at run-time using interpreted equations which were loaded as parameters.

Basic Principles The modelling framework used in these studies was developed with the specific goal of allowing models with arbitrary levels of aggregation and arbitrary temporal and spatial scales to coexist and interact in a model ensemble.

The salient features required are:

- models (and hence any agent) should be able to dynamically set their own state and their local environment,
- models should act in their natural spatio-temporal domain,
- interactions occur in a common spatio-temporal model space.

Emergence There are many potentially emergent properties of the socio-ecological system captured in the Gascoyne and Pilbara InVitro models. For example: seasonally shifting ecological community structure; the evolution of services and industry mixes, regional prosperity, urban development and levels of regulatory intervention.

Adaptation Many of the model components are written to exhibit adaptive behaviour either with respect to their environment or to the prevailing dynamics in the system. This adaptive behaviour was mechanistic and was based on attempting to meet acceptable levels of (rather than maximise) objectives using their existing (imperfect) knowledge of the system and any available (also imperfect) information sources.

Objectives Different models use a range of objective functions. Models which represent animals will use functions which assess the suitability of their environment for breeding, foraging or its "tolerability". The anthropic components have objective functions that are dependent on their level of resolution. For industries represented at the organisational level (petroleum producers) economic returns and the social licence to operate (how popular/unpopular the company is in the community) might be the measures used. Where the actors are at a lower level (individuals or small units) then the assessment measures used to make decisions are defined in terms of income vs. costs (e.g. expected catch vs. costs of fishing those locations), the degree of social network support, or access to recreational or lifestyle "amenities", and experience vs. expectations (conditioned on attitude profiles).

Learning There is no learning amongst the lower trophic levels of the food web, but the highest trophic levels and marine mammals exhibit learning, as do the anthropic model components. The chief expression of this learning is the storage of spatio-temporal response surfaces and calendars that allow for recall of locations, seasons, environmental conditions (etc) where beneficial or adverse conditions were encountered.

Prediction Kalman filters, or simple interpolation of response surfaces, were used to provide a simple predictive ability for the dynamically updating model components. The output of these were used as a predictor of the local state which were then input to fitness functions.

Sensing In principle, the framework imposes no limitations on the abilities of model components to query the state of other components. In practice a detection radius or other filter (e.g. error or decay terms) is imposed dependent on the sensor capability of the requesting agent this is done so that the information available to model components mirrors real world capabilities.

Interaction Most interactions are based on the direct interactions of model components, such as feeding, fleeing, breeding, observing, fishing, booking cruises, utilizing services etc. Some of these interactions are absolute (for example extraction of the hydrocarbon reserve by rigs), while others are conditional or probabilistic (e.g. booking of a room if the budget allows; probabilistic encounter of individual angler with a fish from the local school). Indirect interactions also occur when changes in the state of one component in turn alters the behaviour of another component (such as fin damage by tourists snorkelling through coral gardens leading to higher emigration of reef fish as the reef is degraded; or the demand for labour by the oil and gas industry causing the closure of tourism operations as insufficient staff are available or the displacement of other residents, such as teachers or police personnel, as the costs of living become too high).

Stochasticity Many of the models are stochastic. Ecologically, most use random variables in their movement routines, while the anthropic models include fuzz around the "yes/no" decision point when making a decision and around the result of an event check.

Collectives Within the biophysical components organisms can be represented at a number of scales ranging from single individuals, through superindividuals (schools), to spatially resolved sub-population or domain-wide metapopulation level representations. In the anthropic models individuals may be grouped into family units, tourism groups, or commercial operations.

4.2.4 Details

4.2.4.1 Initialisation

The initial state of the model is defined based on the standing biomasses, spatial distribution and age structures of the ecological components drawn from surveys of the region and records from the local pastoralists (in terms of their livestock holdings). The human demographic structure was taken from ABS Census data, commercial status of the industries from their published annual reports and the zoning and town plans from shire records.

This description is loaded from a number of files which contain the specifications of the selected models and the appropriate parameterisation. Agents can be explicitly listed in the configuration, or they can be started by invoking an agent which generates, in turn, an indicated number of agents of a particular type (with a degree of randomness in some of the state variables of the generated agents, usually the location and possibly biomass). As a general rule, an agent will wait to initialise itself until after the models it depends on have finished initialising.

Initial values for state variables (particularly "parameters") are taken from data collected in the region, published literature or from expert advice. The source of the data is a required attribute in the parameter database. Many parameter values are fixed, but the ability to use expressions as parameters means that parameters can also be set with values drawn from some appropriate random distribution.

4.2.4.2 Input Data

A model in InVitro is defined by a configuration file and a parameter database. These files tell it what models to instantiate and how the models should initialise.

Models are responsible for organising their own sources of data—typically files of some sort—though many use other models as providers of data[1]. If an input data set is not already in the global model domain, remapping its ordinates is usually the responsibility of the model making use of the data. This way, we can carry out computation in the most efficient form, and only project data into a less efficient domain when there are interactions between models.

[1] Recall that environmental data are represented as an agent rather than as an explicit data-element. This approach decouples the models from the details of the implementation, and allows us to replace simple forcing variables with dynamic, interacting models without altering the other models in the ensemble.

Typical data for the various human sectors would include:

- **Fishing** Past catch history (in this case from the Department of Fisheries Western Australia, resolved to a species level and reported spatially), vessel and gear characteristics (supplied by the fisheries operators), prices of species (from Sydney fish markets), price of fuel (time series from ABARES), constraints on site selection, allowed catch and gear characteristics (based on management zones and regulations defined by Department of Fisheries Western Australia or the specific management strategy to be tested).
- **Oil and gas production** global prices, current cost of production, constraints on production/exploration, past production (all time series taken from the annual reports of the oil and gas industry operating in the region).
- **Tourism** locations of accommodation, activities, transport connections (roads, airports, etc), seasonal adjustments, and the characteristics of different types of tourists (taken from maps of the region or from expert reports on the sector in the region, such as Jones et al. (2010)).
- **Pastoral production** attitude profiles for individual pastoralists, initial state of stock and pastures, initial economic state, land use practices, types of investment and maintenance schemes, plans for future development (taken from interviews with the individual pastoralists).
- **Other industries and infrastructure** available workforce and consumer base, value of industry, land use, initial zoning, initial economic state, initial mix of different industrial sectors (essential services, shops, manufacture, etc). These were defined based on documents from the ABS and the local shires.
- **Transport** terminus locations, transport networks (shipping lines or roads, as appropriate), vessel and road train (truck) inventory (from maps of the region or information from the transport companies and the Department of Transport WA).

4.2.5 Conception and Parameterisation of Human Agency

The size and complexity of the InVitro applications makes them effectively equivalent to many ABMs combined. This means they do not sit easily in any one case from the classification in Chap. 1. For instance, in some cases the population size (e.g. number of pastoralists) was small enough that all could be directly interviewed and no upscaling was required; in contrast, upscaling was required to translate the survey data on tourism to span the 200,000+ visitors that enter the region annually.

The rest of this section will combine a brief description of the general form of the anthropogenic components of the InVitro models and how they were parameterised. It would be quite an extensive exercise to detail the explicit application of the framework in each case. Thus a schematic diagram (explained in Fig. 4.3) will be used to summarise the suite of parameterisation options used in M1–M5 for each of the specific anthropogenic components.

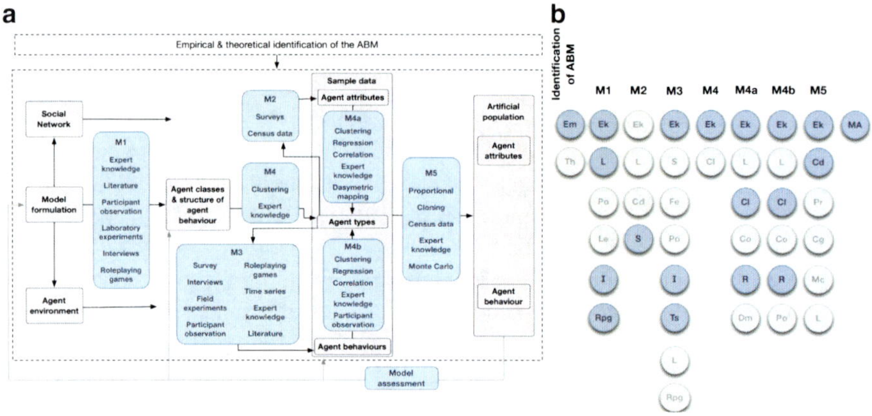

Figure 4.3 Diagram showing how the framework from Chap. 1 (**a**) has been summarised in icon form (**b**). When used to summarise a model in this chapter, methods used to define and parameterise that model will be coloured *blue*, while those not used will be *greyed* out. With *Ek* expert knowledge, *Cd* census data, *Cg* cloning, *Cl* clustering, *Co* correlation, *Dm* dasymetric mapping, *Em* other empirical processing methods (data processing methods, not elicitation methods, i.e. standard clustering methods, conceptual and qualitative models, networks as egonets, regressions and fitted curves), *Fe* field experiments, *I* interviews, *L* literature, *Le* lab experiments, *Mc* Monte Carlo, *MA* model assessment, *Pr* proportional, *Po* participant observation, *R* regression, *Rpg* role-playing game, *S* surveys, *Th* theoretical, *Ts* time series

4.2.5.1 General Form

The studies approached the simulation of human agency (both in terms of decision making and action) in the same way. Decisions would be based on information coming from either direct (perfect) knowledge of the state of the system or, more usually, from another model of information gathering in the domain. The consequent change in the activities affected by a decision were modelled in a way which followed, as closely as possible, the processes and relations in the real world.

There is a significant difference between modelling management bodies and modelling commercial fishers or tourists. Management models can often be constructed to follow real guidelines and regulation, both of which are typically defined explicitly and are available for public inspection. Modelling individuals subject to these regulations (tourists, pastoralists, and fishers) is much more complex, since many things that influence their behaviour are not codified, they may be incapable or reluctant to pass on a deep understanding of these motivations, and their response to novel regulations may be highly uncertain.

In terms of the framework methods used, model identification was done in a participatory way, bringing together theoretical and empirical knowledge. Model characterisation (M1) brought together all possible types of methods, with the anthropic models based either on existing models of the sectors re-implemented (or modified) from the literature or developed based on observational data (interviews, workshops and surveys) collected from the region being modelled. Mechanistic models were implemented in both cases, thanks in no small way to the development of a rapport between the modellers and the local operators which facilitated a very effective use of expert knowledge, participant observation, interviews, surveys, focus groups and meetings with industry experts, regulatory bodies and community members[2]. In addition, laboratory, field, Australian census and market time series were analysed and combined with information from technical and annual financial reports to flesh out model details. As mentioned above, many aspects of InVitro are stochastic and Monte Carlo simulations are the standard means of using the models. No method (from M1–M5 through to model assessment) was used independently. The highly interconnected form of the system model was reflected in the way the framework methods were iteratively used to inform each other improving model performance during hindcast testing of the models (where the model was run for historical periods to see if it could reproduce how the system had behaved through that period).

4.2.5.2 Environmental Protection Agency (EPA)

The EPA model is a regulatory model that acts to modify industry activities based on observed pollutant levels (e.g. in the water column or the tissues of biota like

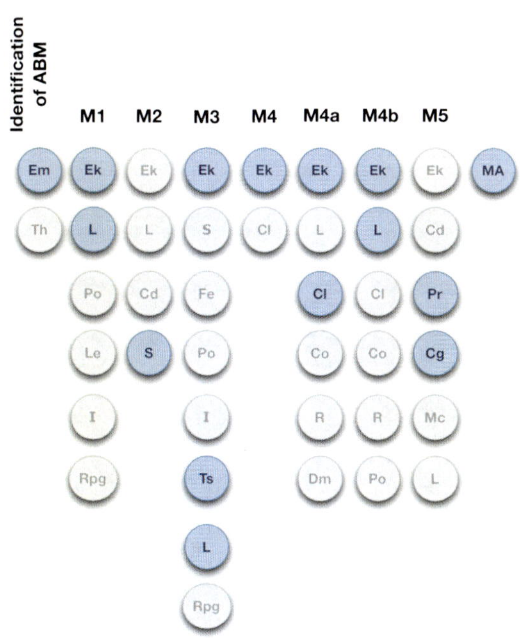

[2] Observations from these meetings influenced all aspects of the models and what they contained. In some cases they were a major source of information (identified in the following sections), in other cases incidental observations helped identify missing or misrepresented components or sources of information that proved useful in detailing the components discussed below (to save space many of these are not detailed below).

prawns). Trigger points are defined in the industry models based on ANZECC guidelines (Australian and New Zealand Environment Conservation Council 2000) and a survey of extra restrictions self imposed by the emitters. When monitored values exceed these levels, fishing activity is spatially restricted, outflows of contaminants (and other polluting activities) are wound down, or other penalties applied. The EPA model acts in concert with other management models to moderate the effects of resource exploitation. Management strategies were implemented as adjustments to the trigger points, variants of the monitoring operations (sensor sensitivity and placement) and via the required responses of the various sectors (speed of closure, form of reductions, period before activities reset once levels drop below the trigger points).

4.2.5.3 Bitterns Release and Other Contaminants

Salt is produced in the region by allowing seawater into shallow ponds and letting the majority of the water evaporate. At some point, there is a sufficient quantity of crystalline salt, and the surplus liquid is flushed out into the coastal waters. This plume solution is both very salty and quite warm compared to the waters along the coast and can cause mortality both through thermal shock, hypersalinity and metal contamination. A simple contaminant mortality model was applied to all biological agents coming in contact with the advected plume. The mortality model in the Pilbara study was a simple piecewise linear model relating dose to mortality in

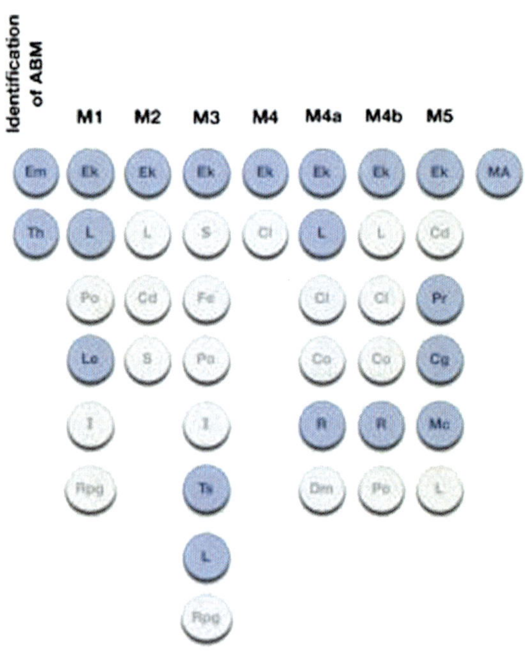

super-individuals. In contrast, the uptake-depuration model used in the Gascoyne study, usually an ordinary differential equation, was specified as an equation taken from the parameter file and resolved using an interpretive evaluator. Both models are able to handle multi-contaminant contact with the assumption that the contaminants are neither potentiating or ameliorating. The parameterisation of the models was a mix of literature values (e.g. for LD_{50} levels) experimental data and biochemical expert knowledge. Movement of the plumes were either explicitly handled within the model (for the Pilbara) or defined by footprints read in from a connectivity model (for the Gascoyne using the Connle model output available at http://www.csiro.au/connie2/). Management of contaminant releases was via the EPA model.

4.2.5.4 Oil and Gas Sector

The oil and gas sector can be represented in two ways. The first is a simple production time series, read in from an external file that then dictates the volume of shipping activity in the local ports (provided by WA Department of Transport). The second uses the rig level production, aggregate demand and technology aspects of the Chi et al. (2009) model. This representation is used to determine the required workforce (scaled based on production relative to the current workforce: production ratio). Reserve depletion and field production are taken from the Moroney and Berg (1999) model. Exploration has been omitted as the northwest Australian fields are already active with defined leases. Organisation specific parameters are taken from

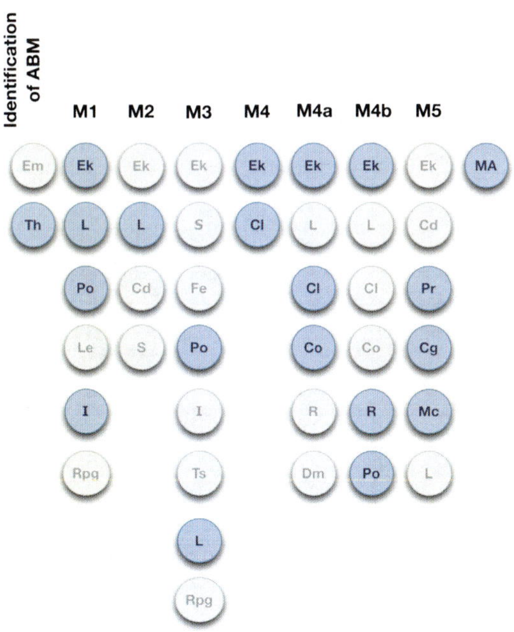

the 2010 annual reports of Woodside, Chevron, Apache, BHP Biliton and Shell. In both cases catastrophic events can be scheduled to occur in a scenario (to examine the implications of a major spill) or they can occur dynamically if simulated ships collide during day-to-day port operations (the frequency of such events can be modified by management decisions to add additional shipping lanes, which can require dredging operations). These plumes are treated using the same mechanics as for the bitterns and pollutants discussed above. Very little of the data required to correctly model the effects of the particular contaminants on the species was available; often the lethal levels were inferred from laboratory data on other species and other age groups.

4.2.5.5 Fishing

Fishing forms a major portion of both the Pilbara and Gascoyne models. It plays a significant social role in both regions and must factor into the management of the regions. Four forms of fishing are considered in InVitro: recreational angling, charter boat fishing, commercial operations (for fin fish and prawns) and mariculture.

Recreational Angling Recreational angling can be modelled in two ways either by an equation-based "catch probability field" or a harvest which is dynamically based on the size of the resident population and the number and distribution of tourists who want to participate in this type of fishing (as opposed to chartered fishing). When using a catch probability field the rate at a location is defined to be proportional to

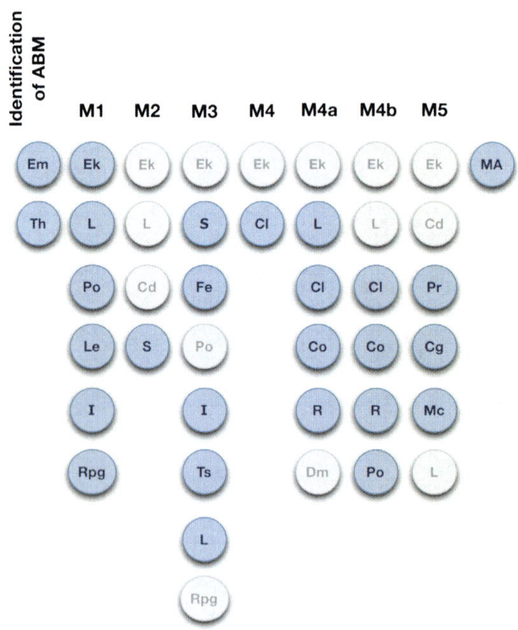

the human population in each settlement attenuated by the distance between the fished location and the boat access points and from there to the settlement(s) via the road network. The tunable parameters, which influenced the basic mortality for recreational catch, were the success rate of the fishers and the travel attenuation values. The success rate was estimated from creel surveys and anecdotal evidence. Travel attenuation was based on fuel prices and a tolerance factor calibrated so that model values matched those from fisheries reports and creel surveys.

In the dynamic angling model, the extent of recreational fishing is closely linked to the model of tourism and implemented as a set of models which simulate the actions of individuals rather than as a "field" effect. This representation was used as evidence from creel and tourism surveys (Jones et al. 2010, Department of Fisheries 2007/2008 survey unpublished data) indicated that the angling behaviour was not only based on access but tourist preferences, budgets habits and available information and gear. It was also significantly more nuanced and adaptive than the simple catch probability field. The number of recreational anglers per half day time-step is defined by the tourism model based on the kinds of tourists in the model and their propensity for recreational fishing (note that one sector of the tourism model is composed of local residents recreating), their location relative to their accommodation (and travel time to both the accommodation and activity site), their budget, available gear, kind of access point and the tourist's desired fishing experience. These factors also dictated whether the angler fished from shore, a small boat in the lagoon or a larger boat (privately owned or chartered) that could travel further afield. Once a location (a rough geographic area) had been chosen a specific coordinate to be fished would be selected. Compliance was not compulsory (a risk trade-off could be made weighing potential returns against risk of detection), but if the fisher complied the fishing location was chosen to avoid incursion into areas where fishing operations were prohibited, such as areas adjacent to pipelines, oil rigs and areas outside the permitted fishing zones.

The patterns of behaviour per class of tourist was parameterised based on extensive survey data collected by Jones et al. (2010), fine scale observations by Smallwood et al. (2010) and participatory activities (e.g. focus groups, interviews and role playing at meetings held in Perth and in the region centres). A Kalman filter and response fields were used to store knowledge of catches and share information between anglers to update expectations and potential fishing locations.

The actual capture of fish and the application of fishing mortality is calculated first conditional on accessibility of the fish, encounter with the gear and the catchability of the fish based on age and selectivity of the gear based on size (parameterised from fishing models for the same gear in other tropical systems (Little et al. 2010) and data collected locally by (Babcock et al. 2009)). The calibration for this model was performed so that modelled values matched those from fisheries reports and creel surveys.

The management of recreational fishing includes size limits, discard rates, prohibited species and protected areas. These constraints were constructed either as a part of a management strategy under trial or were taken from existing environmental, conservation and fisheries regulation. Species or areas could be prohibited as a management response to declining species numbers or other threats to the integrity of the breeding stock, again defined using scenario specific decision rules given in Little et al. (2010).

Charter Boats Charter boats are a special case of recreational fisher that is based on tourism boats. The commercial operations are taken from the tour boat model with the fishing activity taken from the recreational fishing model. Parameterisation of species targeted, cost structures, catch rates and regulations were taken from charter boat information from DoFWA; other behavioural aspects based on interviews, surveys, field data collection and role playing events.

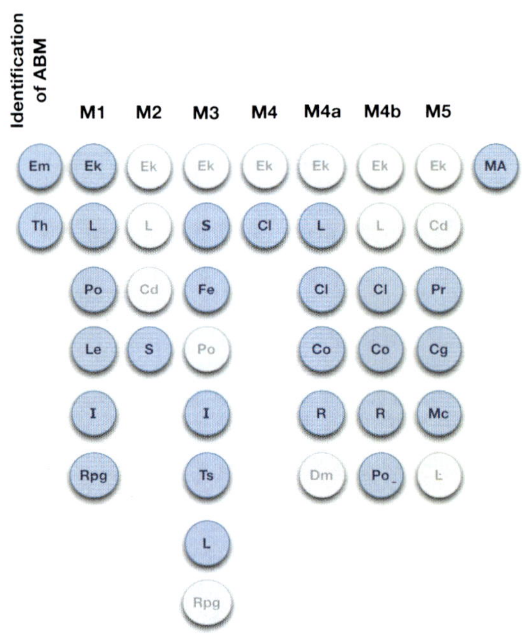

Commercial Fishing The representation of commercial fishing simulates the physical process of steaming to locations, deploying and retrieving gear and returning to port at an individual vessel level. Site selection is based on potential target species (the identity of which could be updated according to the costs associated with the target and the expected value of the catch given market prices), fishing gear available to the boat, the expected value of a catch from a nominated site, and the range of the vessel. The knowledge of the kinds of catches expected at different fishing sites is initially based on historical successes and failures, but is then updated through the course of the simulation (using a Kalman filter) based on realised catches (and information from spotter planes in the case of prawn trawling). Compliance and fine scale site selection is as for the dynamic form of recreational fishing.

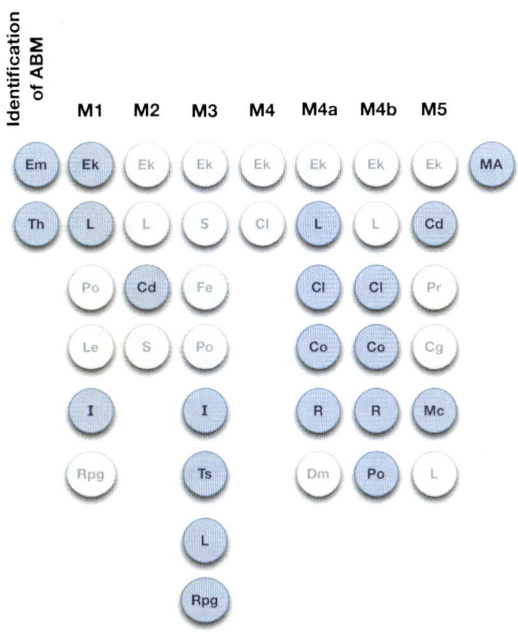

The initial behaviour of each vessel is based on voluntary log book data concerning the fishing operations: trawl location, CPUE for various species and grid references for the cells. These data are seeded into the appropriate Kalman filters in the grids and updated as the simulation progresses. Using a Kalman filter makes it possible for small transient drops or spikes in catches to be accommodated without erasing the historical utility of the location in one step. Individual variability in risk taking is controlled by giving each vessel a greater or lesser willingness to try locations where the uncertainty is high. While it was possible to parameterise in detail the smaller fleets of the northern Gascoyne, Pilbara finfish and prawn fisheries, the larger commercial sector in the southern Gascoyne was parameterised based on interviews and data from DofWA and observations from role playing games. In addition the Monte Carlo handling of the behaviour of the individual fishing boats in the simulation is chosen to span the range between "risk-averse" and "willing to take a chance" in the choice of the location for fishing, though all decisions are filtered through a comparison of the expected yield versus the cost of fishing at the location. This approach means that locations which are historically reasonable places to fish will be considered as options by some fishers even though the estimated error in the expected CPUE was high.

Mariculture Mariculture operations (mainly pearl oysters in the Pilbara) are represented using simple logistic production models coupled with the effects of contaminant plumes. The logistic model was parameterised using data from the company's annual reports and information from DoFWA.

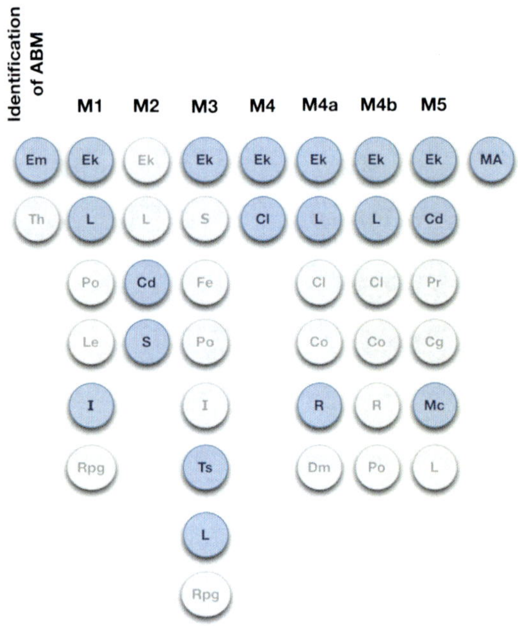

Fisheries Management The fisheries management model monitored biomass, habitat state and aggregate catch reports from vessels. The main management levers are spatial management, seasonal closures, catch and effort quotas (including bag limits) and gear limitations (like mesh or hook size). Management actions are taken in response to an assessment of the status of fish populations against prescribed criteria (in prawn fisheries, for example, the ad hoc opening and closing of fishing zones can be triggered based on acceptable catch rates, with rates defined from historical data so they match the timing of real historical decisions). In the case of recreational fishing this can be socially defined criteria (e.g. likelihood of capturing a trophy fish >40 cm in length in a day's fishing). The management strategies and assessment rules applied are specific to each of the studies (as the mix of management levers and target species differed between the regions) and are either based on historical management documents (e.g. "Pilbara Fish Trawl Fishery (Interim) Management Plan" (1997) and the "Pilbara Trap Management Plan" (1992)) or devised in consultation with DoFWA. Reported catch and effort data from the simulation was used as the basis of simulated management decisions in both studies, and in the Pilbara study this included simulated survey fishing analogous to research trawls as an additional source of input. Note that the true state of the population is not available to the assessment models, with sampling processes analogous to those used to garner data in the real world employed.

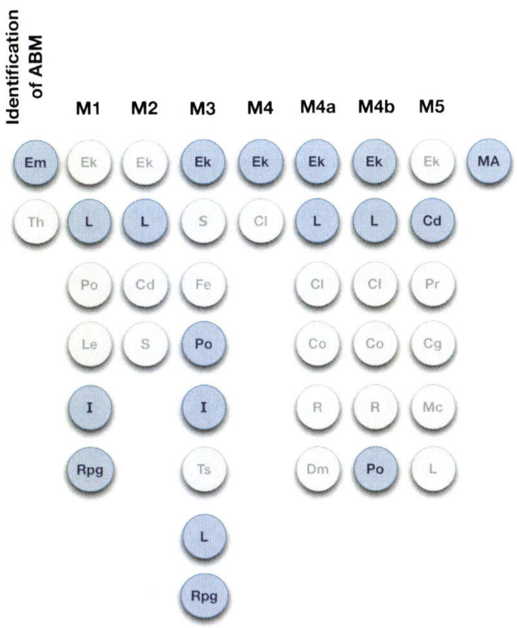

4.2.5.6 Urban Economy, Industry and Development

Land use and services in the urban and suburban areas are simulated with a cellular automaton (CA) based primarily on White et al. (2000) and the demands by developers, households and industry taken from URBANSIM by Waddell et al. (2003)—parameterised based on surveys of local shop keeps and interviews with shire councils and the chambers of commerce in the region. The CA was dynamically supplemented by demand from the tourism, tourism management and economy models (the state of the services in the automata are also fed back to the economy model). In addition, the development of urban sprawl were taken from Leao et al. (2001), planning processes were defined based on discussions with the local shires and the model of Ligtenberg et al. (2001) and the handling of amenities, waste and recycling was based on information from the operators as well as Jones et al. (2010) and Dyson and Chang (2005). Finally, the individual actors in the demographic components of the automata are updated (effectively acting as a nested agent model with dynamic behaviour, activities and political decision making) based on an empirically derived age-structured model using rates from ABS census data. Social attitudes of these “residents” are set using an empirical relationship derived from survey data provided by Jones et al. (2010).

The cellular automaton tracks changes in land use and zoning, employment, social infrastructure, building occupancy and use, and the status and demand on fixed infrastructure such as roads, sewerage, hospitals and schools. This model also feeds services and demand into the models for agriculture and the tourism sector. The initial state of the urban model and its parameters were defined from the council records of the local shires, fine scale aerial photos of the settlements, ABS Data and from parameter values supplied in Jones et al. (2010).

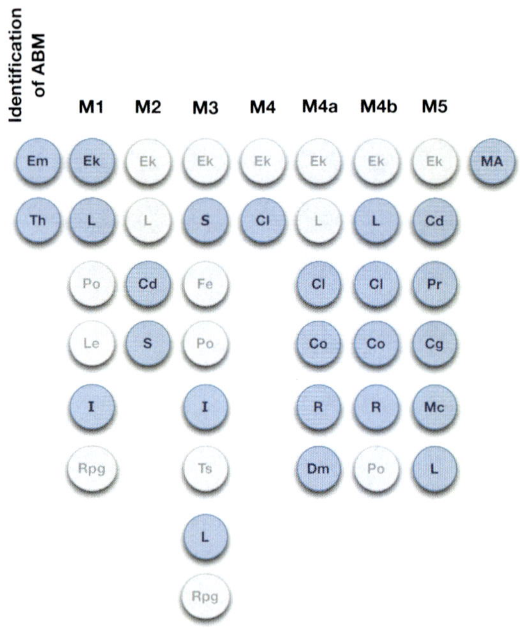

4.2.5.7 Tourism

The tourism industry is a significant contributor to the economic welfare of local communities, and is much more sensitive to the state of the local environment than the mining and petroleum industries. Events that have deleterious effects on the value as a tourist location can influence both the duration of a visitor's stay and their choice of activity.

The underlying structure of the tourism model is largely taken from a tourism destination model (Jones et al. 2010), which was based on data from a repeated survey of tourists in the study domain. This basic behavioural model is coupled to a "management" model that generates the tourist load and allocates "activities" and accommodation to tourists on a daily basis per tourism node (locations pre-defined as major settlement or potential activity sites at initialisation see the sites marked with stars in Fig. 4.1). Each node has a dynamic activity profile which gives a weight to each activity to indicate the quality of the activity at that location. These

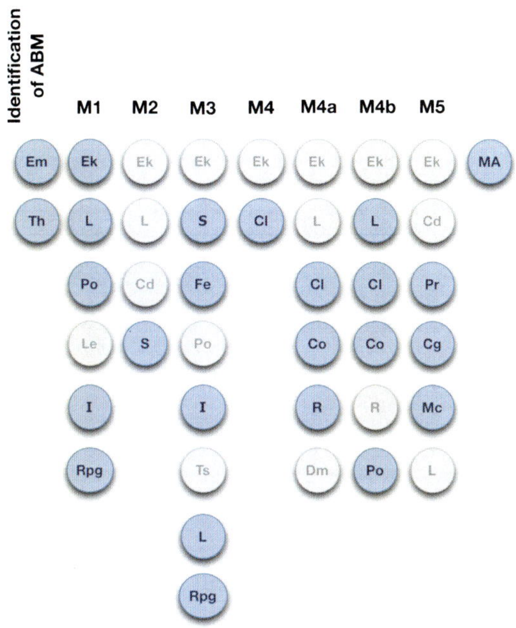

activity weightings are used by the tourists to find locations that best match their desired activity profiles, also taking into account accommodation options and the cost of travel. The nodes are linked by a distance matrix which are travel distances (in kilometres).

The surveys by Jones et al. (2010) identified three general classes of tourist, and these classes were further subdivided on origin (local residents, intrastate, interstate or overseas) and used to assign preferences for accommodation type, preferred activities, group size, trip duration, travel speed, budgets and tolerances (see Table 4.1 for lists of the kind of accommodation and activities available across the nodes). Additional data (such as prices for different types of accommodation) were garnered from advertisements, census data, and Northcote and Macbeth (2008).

The individual behaviour of the tourists is based on concepts from Laporte and Martello (1990), and Gimblett et al. (2003). The decision algorithm used is described in Algorithm 4.2.

```
begin
  begin
    Store output statistics. Check for any tourists exiting the region.
    Check for new tourists entering the region.
  end
  begin
    Do active tourist's currently scheduled activity. Identify recfishers
    (so have effort for recfishing model).
    Decide on next activity.
    Decide on location of activity:
    select best location in location list from
      Calculate score on the correspondence between its activities
      and desired activities, with travel costs, distance,
      accommodation costs, accommodation score, whether there
      will be time to do the activity before dusk and the node's score
    end
  end
  Do Tourist night activities (once day's worth of activities and travel are
  complete update satisfaction and knowledge).
  Set the agent's subjective time
  Re-insert the agent into the run-queue and increase its age (trip
  length).
end
```

Algorithm 4.2 Steps in the tourism model

Table 4.1 Tourist accommodation and activities

Accommodation[a]	Activities[b]	
Camping	Beach leisure	Shopping
Caravan Park Backpackers hotels	Shore based angling	Eating out
Rent	Charter boat fishing	Sight seeing
Other	Private cruiser fishing	Safari (on land)
	Small boat fishing	Marine wild life tour
	Snorkelling	Surfing
	Scuba	Other

[a] Each accommodation type at each location has a cost per night per person and the number of available beds (which cannot be exceeded and is updated whenever tourists vacate, or move into, an accommodation)

[b] At the start of the tourist's trip the total number of activity hours is calculated (based on assuming a certain number of active hours per day). Each activity takes different amounts of time and the tourist's activity preferences are assumed to indicate the proportion of their total activity hours they wish to spend doing each activity. When an activity is done the proportion of their total activity hours is calculated and the activity preferences are updated to reflect the tourist fully or partially 'ticking off' this activity from their list. Each tourist has an individual (imperfect) knowledge of the activities that are available at each location, which can be updated through education, advertising or social networking

The tourism model interacts with the economy model to set tourism values, pick up wage rates, and affect a number of sectors in the economy (retail, hospitals, services, etc). It also interacts with the agriculture model since many tourism nodes are on pastoral stations which can influence the accommodation and activities available. The tourism model also interacts with the urban model (which provides accommodation and services), the charter vessel and recreational angling models and the tourism operator model.

4.2.5.8 Agriculture

The agricultural sector is modelled as a hierarchical collection of entities, namely farmers, pastures, and livestock. A farmer agent manages a number of pastures and herds of livestock (such as cattle, goats or sheep), employs staff and makes investment decisions, such as the amount and type of tourist accommodation to build and maintain. The agricultural model interacts with the terrestrial environment (e.g. the pasture vegetative growth) and the tourism model (by providing accommodation and activities for the node). Farms provide impetus to the economy by direct production of livestock, jobs and services associated with tourism. Livestock and tourism can have direct effects on the underlying ecosystem model through effects on vegetation and erosion, for example.

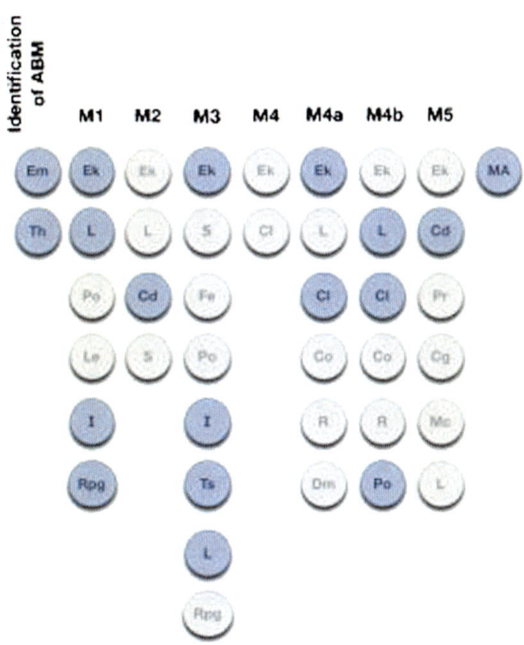

After long discussions with pastoralists in the region about their properties and how they managed them and a review of existing pastoral models it was clear that existing models of Australian range-lands (with minor modifications to deal with aridity) would be suitable models of the system. Consequently, the basic structure of the range-lands pastoral model was taken from the SEPIA model implemented originally for the Bowen-Broken catchment in Queensland (Smajgl et al. 2007). This was supplemented by a livestock model created by combining the equations for sheep and pasture given in Janssen et al. (2000) and Cacho et al. (1995) with the arid grazing and goat models given in Richardson et al. (2005), and Sparrow et al. (1997). Lastly the investment model was based on the innovation and information sharing model given in Berger (2001).

Data about farming practices, stocking levels and the interactions between farmers and other parts of the social and economic milieu were obtained by interview. Talking with pastoralists made it possible to rank their priorities and to tease out their connections with the rest of the local economy and their role in the social matrix. The final parameterisations (particularly of the attitude profiles) were calibrated so that the emergent production statistics and investment patterns matched historical records.

4.2.5.9 Transport

The shipping and road train models were empirical, mechanistic descriptions of the activities of these sectors, i.e. numbers of ships/road trains required per tonne of product (either produced locally or required—all goods must be shipped in), constrained by available pens in ports or available road trains and driver accommodation

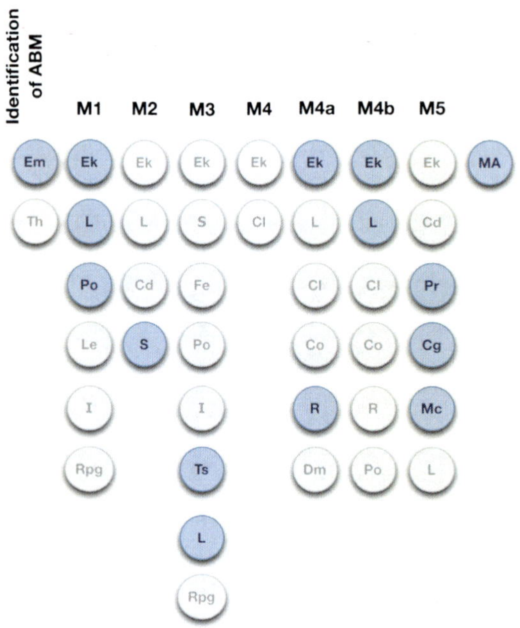

(initialised based on reports from the LinFox transport company which is the major supplier in the region). Ships were constrained to travel along marked shipping lanes and road trains along paved roads. In the case of the road trains the trucks must cycle through maintenance based on standard services or if they were involved in an accident (e.g. significant interaction with livestock, macropods or other road vehicle). Collisions (either between ships or between road vehicles) were set based on rates taken from insurance company reports for the regions. The transport model interacted with the economy model to report employment sector activity rates and to pick up wage rates.

4.2.5.10 Economy

The regional economy is represented by an annual implementation of a standard input-output economy model. This model is initially populated from regional state treasury analyses for the major settlements in the Gascoyne and is then updated dynamically based on the activities if the various sectors in the model ensemble. The resulting wage rates and costs of living are fed back to the sector and demographic models to set the cost structure in the net year of the simulation.

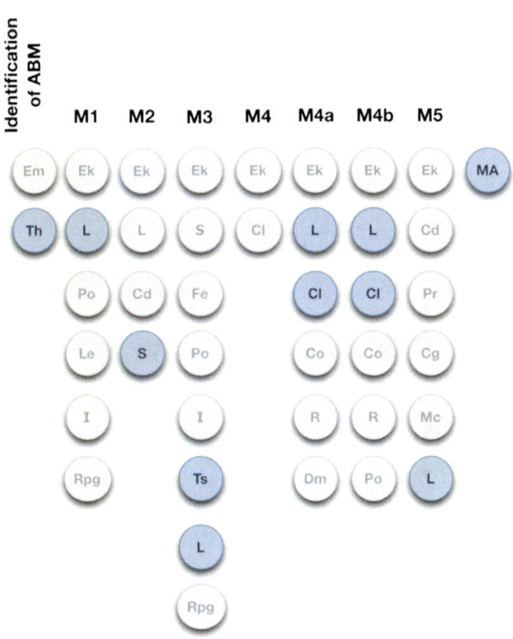

4.3 Discussion

Both of these InVitro models stand both on the shoulders of their antecedents and the participatory information provided by stakeholders in the region and many regulatory bodies. Many of the models represented in the ensembles are well-established models in their own right that have had significant testing through time, and span a range of representational forms, spatial and temporal scales and roles. Building on these component models involves creating linkages between them: a process which may involve a considerable amount of code to manage interactions between models which have different views of the granularity of space or time, but which also involves calling on the full suite of model parameterisation methods included in the framework discussed in Smajgl et al. (2011).

Building such a complex model is quite challenging both computationally and in terms of populating it, interpreting and communicating it to the many potentially interested parties. Participatory methods have proved particularly useful in providing the difficult to collect behavioural information and increasing understanding of the models contents and dynamics. In this context, it has been worth keeping InVitro versions of models as close to their "canonical form" as possible. This makes it easier to prune back to the bare bones of the model, restricting interactions with other models and *gradually* increasing the complexity of the system. Many useful models are developed in isolation and coupling them to others introduces dynamics which may not be well catered for in the original form; the process of gradually introducing interactions makes it easier to ensure that the coupled dynamics don't undermine the robustness of the model, but demands a great deal of flexibility on the part of the embedding framework. Importantly though, it naturally facilitates using a broad range of methods to parameterise and test the model forms. Every type of method listed in Smajgl et al. (2011) was used somewhere in the final full system model (e.g. expert knowledge, literature, surveys, interviews, field collections etc). However, not every method was appropriate in every case; and just as the form of the models of each part of the system was tailored to what best represented that system, so the methods used to parameterise each part should be selected based on what will best supply the needed information given the characteristics of the system in question (e.g. what works well for a very small local community cannot be blindly be applied to collect information on large transitory populations that move through the same geographic location).

On both of the InVitro applications discussed here, the models of human agency largely began with existing models as a starting point, but coupling this work with active engagement with local stakeholders was a critical part of the development process. Detailed conversations with pastoralists, tourism operators, local residents, council members and others were used, along with the integrated research undertaken by members of the Ningaloo Research Cluster, to provide a richer picture of local activities with an improved level of detail.

When dealing with processes on so many scales the value of existing reports and theory cannot be denied (see the many cases here where the models began using skeletons sketched based on studies undertaken elsewhere). Similarly, statistical tools have a key role to play as the many data sources are brought together. Such methods are particularly important when dealing with sensitive information (e.g. commercial, social or cultural data) that must be processed before being used in a public forum. Other methods (such as pattern oriented modelling) are also incredibly useful when testing models in hindcast modes or against "test sets" (data and conditions not used during the original model formulation and training).

Nevertheless in ABMs which are heavily dependent on the action of human components there is a critical need to collect information on decision-making and behaviours that are not well quantified (or in some cases described) in the literature or statistically. This was certainly the case in the InVitro modelling exercises discussed here, which would have floundered (particularly in the highly interconnected Gascoyne case) without a heavy use of participatory methods. In this work the form of consultation, the approachability of the scientists and their ability to engage with their audience has been important to success. Humans are not necessarily consistent and don't necessarily behave in an optimal manner. There are often important data that affect the effectiveness of the modelling of human interactions amongst the socio-economic sectors, and with the ecosystem, that can only be gleaned by listening to people actually living in the system; people who also trust the scientists developing the models and the utility of MSE sufficiently to share their knowledge and experience. This not only benefits model construction, but also sees a much-improved likelihood of insights drawn from the model being used as intended. Lastly, it significantly increases the likelihood of a productive on-going relationship between the model developers and potential users. While highly beneficial, this level of interaction can be a costly exercise (both in terms of time, good will and financially if working with a location remote from research centres). The magnitude of the problem grows with the size of the population involved and the studies would not have been as successful without embedding researchers in the local community and the regulatory bodies for months at a time.

References

Australian Bureau of Statistics (ABS) (2012) State and territory statistical indicators. Technical Report 1367.0, Australian Bureau of Statistics

Australian and New Zealand Environment Conservation Council (ANZECC) (2000) Australian and New Zealand guidelines for fresh and marine water quality. Volume 1: The guidelines. Australian and New Zealand Environment and Conservation Council, Agriculture and Resource Management Council of Australia and New Zealand

Babcock R, Haywood M, Vanderklift M, Murphy N, Clapin G, Limbourn A, Kozak D, Thomson D, McCleod I, Milton D, Skewes T, Dennis D, Depczynski M, Heyward A, Radford B, Moore J, Fitzpatrick B, Rees M (2009) Lagoon invertebrates final report and monitoring recommendations. WAMSI node 3 project 3.2.2d final report. Western Australia Marine Science Institute

Berger T (2001) Agent-based spatial models applied to agriculture: a simulation tool for technology diffusion, resource use changes and policy analysis. Agric Econ 25:245–260

Blaber SJM, Cyrus DP, Albaret J-J, Ching CV, Day JW, Elliott M, Fonseca MS, Hoss DE, Orensanz J, Potter IC, Silvert W (2000) Effects of fishing on the structure and functioning of estuarine and nearshore ecosystems. ICES J Mar Sci 57:590–602

Butterworth DS, Punt AE (1999) Experiences in the evaluation and implementation of management procedures. ICES J Mar Sci 56:985–998

Cacho OJ, Finlayson JD, Bywater AC (1995) A simulation model of grazing sheep: II. Whole farm model. Agric Syst 48:27–50

Chi KC, Nuttall WJ, Reiner DM (2009) Dynamics of the UK natural gas industry: system dynamics modelling and long-term energy policy analysis. Technol Forecast Soc Change 76:339–357

de la Mare WK (1998) Tidier fisheries management requires a new mop (management oriented paradigm). Rev Fish Biol Fish 8:349–356

Dyson B, Chang NB (2005) Forecasting municipal solid waste generation in a fast-growing urban region with system dynamics modeling. Waste Manag 25:669–679

Francis RICC, Shotton R (1997) Risk in fisheries management: a review. Can J Fish Aquat Sci 54:1699–1715

Fulton EA, McDonald AD, Hayes D, Lyne V, Little LR, Gray R, Fuller M, Condie S, Scott R, Webb H, Hatfield B, Martin M, Sainsbury K (2006) Multiple use management strategy evaluation for the north west shelf specification for Australia's north west shelf. Final report 15, CSIRO Australia, Hobart, Tasmania

Fulton EA, Gray R, Sporcic M, Scott R, Little LR, Hepburn M, Gorton R, Hatfield B, Fuller M, Jones T, de la Mare WK, Boschetti F, Chapman K, Dzidic P, Syme G, Dambacher J, McDonald AD (2011) Ningaloo Collaboration Cluster: adaptive futures for Ningaloo. Technical report 5.3, Ningaloo Collaboration Cluster, Hobart, Tasmania

Gimblett R, Lynch J, Daniel T, Ribes L, Oye G (2003) Deriving artificial models of visitors from dispersed patterns of use in the Sierra Nevada wilderness, California. J Nat Conserv 11:287–296

Gray R, Fulton E, Little R, Scott R (2006) Operating model specification within an agent based framework. Final report 16, CSIRO Australia, Hobart, Tasmania

Grumbine RE (1994) What is ecosystem management? Conserv Biol 8:27–38

Janssen MA, Walker BH, Langridge J, Abel N (2000) An adaptive agent model for analysing co-evolution of management and policies in a complex rangeland system. Ecol Model 131:249–268

Johnson JC, Borgatti SP et al (2003) Network role analysis in the study of food webs: an application of regular role coloration. J Soc Struct 2(3). http://www.cmu.edu/joss/

Jones T, Wood D, Hughes M, Pham T, Pambudi D, Spurr R, Dwyer L, Deery M, Fredline L (2010) Building a sustainable planning tool for Australian tourism destinations. Final report Ningaloo Collaboration Cluster & report for the sustainable tourism CRC. Curtin University of Technology, Centre for Research and Graduate Studies-Humanities Perth, Western Australia

Laporte G, Martello S (1990) The selective travelling salesman problem. Discret Appl Math 26:193–207

Leao S, Bishop I, Evans D (2001) Assessing the demand of solid waste disposal in urban region by urban dynamics modelling in a gis environment. Resour, Conserv Recycl 33:289–313

Ligtenberg A, Bregt AK, van Lammeren E (2001) Multi-actor-based land use modelling: spatial planning using agents. Landsc Urban Plan 56(12):21–33

Little LR, Thebaud O, Boschetti F, McDonald AD, Marriott R, Wise B, Lenanton R (2010). An evaluation of management strategies for line fishing in the Ningaloo Marine Park. Final report, Ningaloo Collaboration Cluster

Milner-Gulland EJ, Arroyo B, Bellard C, Blanchard J, Bunnefeld N, Delibes-Mateos M, Edwards C, Nuno A, Palazy L, Reljic S, Riera P, Skrbinsek T (2010). New directions in management strategy evaluation through cross-fertilization between fisheries science and terrestrial conservation. Biol Lett 6:719–722

Moroney JR, Berg MD (1999) An integrated model of oil production. Energ J 20:105–124

Nellemann C, Hain S, Alder J (2008) In Dead Water: merging of climate change with pollution, over-harvest, and infestations in the world's fishing grounds. Norway, GRID-Arendal

Northcote J, Macbeth J (2008) Threshold of tolerability: the impact of management changes to recreational fishing in Ningaloo Marine Park. Tour Rev 63:28–35

Richardson FD, Hahn BD, Hoffman MT (2005) On the dynamics of grazing systems in the semi-arid succulent karoo: the relevance of equilibrium and non-equilibrium concepts to the sustainability of semi-arid pastoral systems. Ecol Model 187:491–512

Smajgl A, Heckbert S, Bohnet I, Carlin G, Hartcher M, McIvor J (2007) Simulating the grazing system in the Bowen broken: an agent-based modeling approach. Report, CSIRO Australia, Canberra

Smajgl A, Brown DG, Valbuena D, Huigen MGA (2011) Empirical characterisation of agent behaviours in socio-ecological systems. Environ Model Softw 26:837–844

Smallwood CB, Beckley LE, Moore SA, Kobryn HT (2010) Assessing patterns of recreational use in large marine parks: a case study from Ningaloo Marine Park, Australia. Ocean Coast Manag 54:330–340

Sparrow AD, Friedel MH, Stafford Smith DM (1997) A landscape-scale model of shrub and herbage dynamics in central Australia, validated by satellite data. Ecol Model 97:197–216

Waddell P, Borning A, Noth M, Freier N, Becke M, Ulfarsson G (2003) Microsimulation of urban development and location choices: design and implementation of urbansim. Netw Sp Econ 3:43–67

White R, Engelen G, Uljee I, Lavalle C, Ehrlich D (2000) Developing an urban land use simulator for european cities. In: Fullerton K (ed) Proceedings of the 5th EC GIS workshop: GIS of tomorrow. European Commission Joint Research Centre, Brussels, pp 179–190

Chapter 5
Using Spatially Explicit Marketing Data to Build Social Simulations

Andreas Ernst

5.1 Introduction

Agent based modellers often encounter one of the two following problems: Either they can use available large-scale demographic or land-use data of built-up areas to inform the construction of their agent population. Then, however, if the wish or the necessity arises to provide their agents with empirically based data on behavioural motivations, knowledge, and the like, they will be unable to gather the appropriate data on all individuals for the whole area they consider.

In the other case, detailed data on cognitive and behavioural variables have been gathered e.g. through a domain-specific survey. However, only rough algorithms are available to scale up this detailed knowledge to a larger area. Some rely on socio-economic status, others on household size and household composition. In this process, unfortunately, some of the differentiated nature of the gathered data is lost.

The problem we face here is termed the upscaling problem. It can be regarded as a general problem of spatial ABM: Fine grained data relating to the domain and the research question of interest needed to build agents' behavioural rules are just not available on a larger scale—and it would often be quite unrealistic to gather them on that scale. To bring more of a realistic representation of agent behaviour on the larger scale, it would be necessary to have a micro-macro bridge that does not solely rely on structural variables like demography, income or household size, but on concepts more directly related to behaviour.

The sociological classification of individuals into lifestyles, or milieus, can be regarded as such a bridging concept. Lifestyles are the division of population into classes of like-minded people. They do not have to have the same economic, educational or occupational background, but they share what they like and what they buy, which TV programs they watch and which political measures they appreciate. People

A. Ernst (✉)
Center for Environmental Systems Research, University of Kassel,
Wilhelmshöher Allee 47, 34109, Kassel, Germany
e-mail: ernst@cesr.de

A. Smajgl, O. Barreteau (eds.), *Empirical Agent-Based Modelling – Challenges and Solutions*, DOI 10.1007/978-1-4614-6134-0_5,

classified into one lifestyle thus share their fundamental values as well as everyday attitudes towards work, family, leisure, money, or consumption (Bourdieu 1984).

The power of the lifestyle concept comes to bear when data about the residence location of individuals belonging to a certain lifestyle can be used by the modeller. These data are important for so-called geo-marketing purposes and thus are commercially available in one form or the other for most industrialised countries.

This paper describes a generic approach of using such spatially explicit marketing data to build social simulations. Commercially available data are used to inform the agent building process about demographic and economic variables of households, their lifestyles and precise geographical location. The lifestyle classification is used to connect data gathered for specific purposes and the spatial larger scale representation. The resulting models can easily be combined with geo-bio-physical models like e.g. hydrological, agricultural, or meteorological ones to highlight human-environment interaction and diversity in space.

To illustrate the general approach, this paper will draw especially on one agent-based model within the GLOWA-Danube project that aimed at providing an integrated computer model of natural and social processes in the upper Danube catchment in Southern Germany.

5.2 Model Description

5.2.1 Motivation

The motivation for building DANUBIA, the integrated model of the upper Danube region was to provide a useful basis to advance environmental decision making. To this aim, a valid "core engine" integrating all domain relevant processes from the different fields and disciplines and their interactions was conceptualised and implemented. The GLOWA-Danube project was sponsored by the German Ministry of Education and Research from 2000 to 2010, and was the first such enterprise on a regional scale (Mauser 2000; Ernst 2002, 2009). The river basin considered in the model has an extension of approx. 75,000 km^2 ranging from the Alps to the Bavarian lower plains and includes parts of southern Germany, Austria, and Switzerland. About 11 million people are living there, and the basin includes high mountains, agricultural regions as well as big cities such as Munich.

5.2.2 The Modelling Framework

The DANUBIA system integrates 16 fully coupled process models from 11 scientific disciplines ranging from hydrology to environmental psychology and from meteorology to tourism research (Ernst et al. 2005; Soboll et al. 2011; for a description of DANUBIA from a computer science perspective, see Barth et al. 2004). The general objectives, the methodological framework and results of DANUBIA have been published

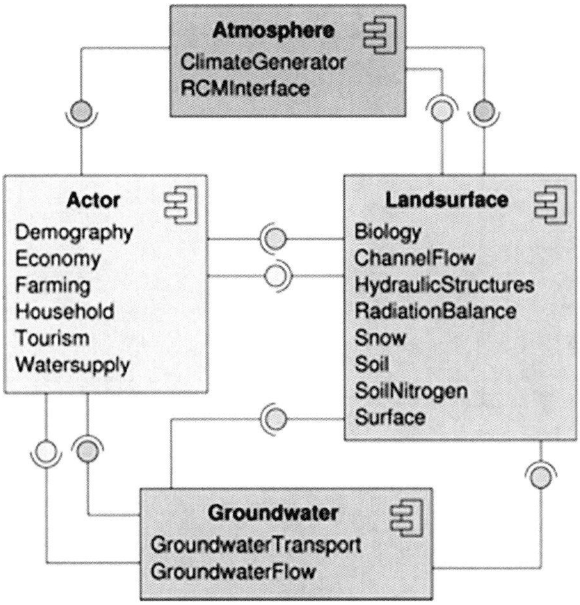

Figure 5.1 *UML diagram of the four DANUBIA components Atmosphere, Landsurface, Groundwater, and Actors*. The components are shown together with the respective models and their interfaces. (From Elbers et al. 2010)

in previous papers (e.g. Hennicker and Ludwig 2006; Barthel et al. 2008). The system structure follows the structure of the domain: There are five components (Landsurface, Atmosphere, Groundwater, Rivernetwork, and Actor) as represented in Fig. 5.1.

Each component encompasses multiple models. For example the actor component, which collects process models from the social sciences, comprises implementations of the Household, Demography, Economy, Farming, WaterSupply, and Tourism models. The components and their models are interconnected via a simulation framework which assures the communication linkage through interfaces, the setup and monitoring of simulation runs, the logging of model states, etc. Agent based modelling plays a central role in the actor component.

In response to the need for a spatial resolution system differentiated enough to enable detailed analyses, a lattice was superimposed on the study area. The spatial representation in DANUBIA is realised using a 1 × 1 km unit, a "proxel" (for "process pixel"). A proxel is the common spatial unit of GLOWA-Danube and is fitted with general attributes like the altitude or the number of inhabitants. All simulated elements, such as vegetation, water supply companies, wells, or all kinds of water users, are located on proxels and all simulation processes are carried out on the proxel level (Kneer et al. 2003).

This unit constitutes a compromise between the various disciplines participating in building the DANUBIA system with regard to the scale and the shape of their spatial representation. While some disciplines have difficulties in downscaling their computations to the 1 × 1 km unit, others have to upscale, and yet others (especially from the behavioural sciences) have to translate from and to spatial representations that are usually oriented towards administrative boundaries. These shapes were mapped to the proxel logic using GIS procedures. From the more than 75,000 proxels representing the size of the Danube river basin, 9,210 are inhabited.

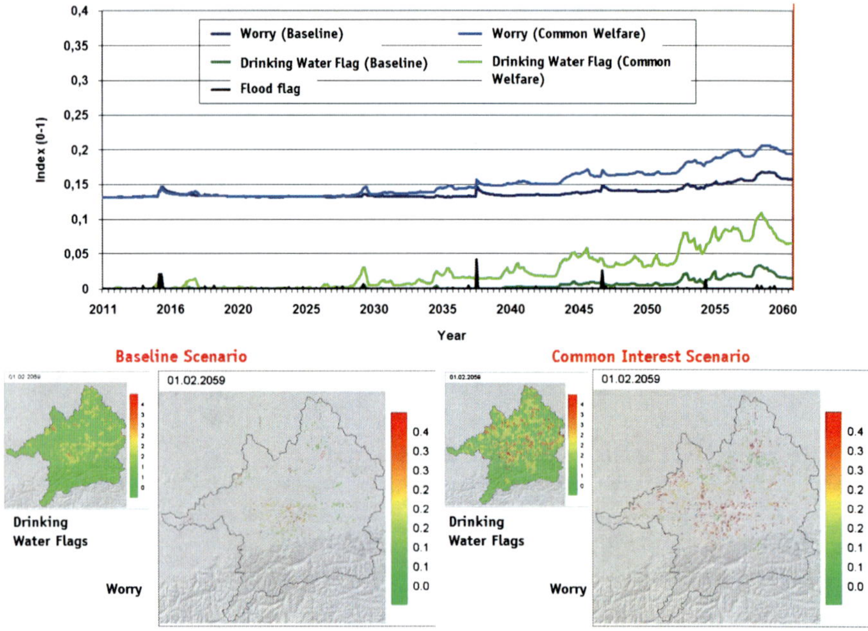

Figure 5.2 *Presentation of the results of a model run of the Household model within DANUBIA.* The comparison of the household's perception of water sustainability in two societal scenarios (baseline vs. common interest) is shown. The scenarios differ in the importance the water suppliers give to the scarcity information (the so-called drinking water flags) they provide to the citizens. Clearly, spatial patterns of households' worry about water sustainability can be identified. They directly relate to the summary *line graphs* on the *upper* half of the figure, which show a slow increase of both scarcity information given and the resulting households' reactions over time

The temporal unification in the DANUBIA system is realised by a so-called time controller (Ludwig et al. 2003). It coordinates correct temporal data exchange between models via interfaces, thereby allowing for a parallel interactive analysis of physical and socioeconomic processes. The models in DANUBIA work on different time steps, however. These range from less than 1 h (e.g. in the forest growth model) to 1 month of model time (e.g. in the actor models). Each model logs on to the time controller for an integrated model run and provides its time step and other relevant details for a causally correct temporal interfacing of all models. The controller is tolerant towards models that are missing or die. In such a case, pre-fabricated output data of the most similar run are taken from a data base, with of course losing all run-time feedbacks with that model for that run.

The development of DANUBIA has been embedded in a policy process, with stakeholders from science, administration, and government accompanying it and giving feedback on the models and the scenarios to be computed. They helped identify relevant topics, provided data and further analysis, and discussed results and their applicability. Especially the graphs and the movies that were used to present the results of the computed scenarios proved a means well fitted to the needs of the stakeholders (see Fig. 5.2). More on the scenarios construction process can be found in Soboll et al. (2011).

In the following, the *Household* model will serve as an example to describe agent based and spatially explicit modelling relating to domestic water use and water related satisfaction in the DANUBIA system, with special attention to the use of lifestyle data.

Additional information on the *Household* and the other models of the DANUBIA system can be found at www.glowa-danube.de. The software is open source and can be downloaded together with example data from the DANUBIA website http://www.glowa-danube.de/de/opendanubia/allgemein.php.

5.2.3 The **Household** *Model*

5.2.4 Purpose

Within DANUBIA, i.e. the integrated model of GLOWA-Danube, the *Household* model has the objective to compute domestic drinking water use (submodel *WaterUse*) and the water related risk perceptions (submodel *RiskPerception*) in the upper Danube catchment. A third sub-model (*InnovationDiffusion*) is responsible for simulating the diffusion of water saving appliances in the households which in turn influences domestic water use.

The output variable drinking water use is computed as a summary value as well as split into ten different water uses. Risk perception is given as the activation or worry of a modelled household about the state of the resource.

5.2.5 Entities, Scales, and State Variables

The *Household* model receives input data from the *Demography*, *Economy*, *WaterSupply*, and *Meteorology* models during run time. Its output is mainly delivered to the *WaterSupply* (water demand) and the *Rivernetwork* components (for waste water), and to the user interface when providing data that are not used as inputs to further model computations but are presented to the end user (like household satisfaction).

The modelled agents are representative households, five on each square kilometre of the modelled area. Altogether, this sums up to 46,050 representative agents. Each of them represents the empirically determined number of households of one of five different lifestyles per square kilometre (see below), a number which may range from a few dozen to more than a thousand households. The quantities resulting from the agents' actions are weighted with that number and with the mean number of people in the representative household.

The agents are characterised by a higher number of state variables. Among them count *ID*, *location*, *age*, *income*, *modernism*, *importance of price*, *importance of peers*, *importance of the natural environment*, a list of peers that the agent is connected to, the appliances available in the household, and others. Moreover, some state variables change over time through learning, e.g. the agent's memory for events in the social and bio-physical environment, its *activation*, among others.

5.2.6 *Result Presentation*

The results from the *Household* model are presented along with the quantitative climate scenarios that drive a run in DANUBIA and information about water availability, water prices, and water scarcity from the *WaterSupply* model. The results concern the spatial as well as the temporal patterns of the output variables domestic water use and water related household satisfaction. They are presented through line graphs and spatial maps of the catchment area. In oral presentations, the spatial development in the maps over time is shown via movies that are accompanied by the respective line graphs (Fig. 5.2).

The scenarios are analysed and compared with regard to individual and collective welfare, stability vs. volatility of the macro-system, the influence of the structure of social networks, their spatial patterns and the influence of lifestyles.

5.3 Overview: Framework Specific Sequence of Model Building

In the description of the sequence of empirical activities that led to the *Household* model, the exposition follows the CAP framework proposed by Smajgl and Barreteau in Chap. 1. To define agent classes and the structure of agent behaviour in each class (M1) was simple for the *Household* model, since only one class of agents, namely households, had to be considered. The structure of the agent behaviour to be considered was also defined by the mission: To describe domestic water use and water related perceptions. To define the agent types (M4), the categorisation into different lifestyles on the basis of a ready-made, commercially available classification was taken as a guiding framework. Though the classification originally provides ten lifestyles, the survey and behavioural data (see below) would not allow discriminating well between all ten classes. So, they were grouped into five so-called lifestyle groups that made up the five agent classes considered in the model.

To inform about agent attributes (M2), a large survey on the appliances in the households, the knowledge related to water, the sensitivity to price changes, among others, were conducted. To inform about water related agent behaviour (M3), the survey furthermore gathered data about water using habits in the households. Expert knowledge was input via expert workshops, with experts from the marketing enterprise responsible for the lifestyle classification and paid through the project, to learn more about the influence of lifestyles on environmental behaviour. Telephone interviews were carried through to know about the willingness to innovate and to purchase water saving appliances for the household.

From agent attributes (M4a) and agent behaviours (M4b), the gathered data could directly be fed back into the agent typology. Since this typology was defined by the lifestyle classification from the beginning, those data merely enriched the existing types and did not modify the typology. Finally, the upscaling process from the five agent types to the artificial population (M5) was driven by the spatial distribution of the lifestyle data, which was provided by another marketing company (for details see below). The spatial distribution relies on a vast set of data about the financial situation of people and is strongly connected to their housing situation. Thus, the total of around 4 million households in the catchment area could be positioned in its 9,210 inhabited square kilometres. All those steps will be presented in more detail in the following section.

5.4 Technical Details

5.4.1 Definition of Agent Classes and the Structure of Agent Behaviour (M1)

Domestic drinking water demand is determined by three factors: (1) The water use behaviour of consumers, i.e. the frequency of different types of water use, (2) the efficiency of installed water use devices and (3) the diffusion of water saving appliances. The following sections will describe the gathering of data for filling in information about agent attributes and agent behaviour in the DANUBIA *Household* model. While data on the technical features of different devices could be extracted from the literature, both water use behaviours and the adoption of different water use technologies vary depending on specific properties of households such as household size, income, lifestyle and the composition of their social networks by which they exchange information and opinions. Moreover, assumptions about the agent architecture to be psychologically plausible as possible have to rely on both literature and data.

The objectives of the *Household* model were clear-cut: to model the quantity of domestic water use and the water related perceptions in a river catchment. In terms of output variables, this seems quite simple. The agent classes are predefined from the outset, and the behaviour to be reconstructed encompasses only a few variables.

The complexity of the model's mission results from (1) the spatial differentiation needed to communicate with the other models in the DANUBIA compound, and (2) from the need of a empirically plausible, and in this case this means a psychologically plausible reconstruction not only of the overt behaviour, but also of non-observable variables like worry resulting from non-sustainable resource use, or fine-grained decision mechanisms that can be used to model the households' reactions to policy measures. Households are treated like one decision-maker in this model. We abstract from the communication and the discussions within a household or family.

Among the psychologically plausible behavioural mechanisms to be modelled, we count habitual behaviour (most of people's daily water use is driven by habits), multi-attribute decision making (e.g. in the case of the acquisition of a more costly water saving household appliance), and heuristic behaviour (which can be readily used in the case of insufficient knowledge). These behavioural mechanisms allow on one hand modelling purchases of water saving appliances to take into account the diffusion of water saving innovations, which is one of the main factors for a decreasing domestic water use in Germany. On the other hand, they are the core for the households' reactions to exceptional events like changes in water price, or heat waves. They are supplemented by a subjective perception component for the processing of scarcity signals from the water supply companies as well as the social influences from the peers in the social network, and a memory to enable an agent's adaptation and learning past events.

5.4.2 Defining Agent Types (M4)

5.4.3 Literature Review and Theoretical Assumptions

The *Household* model relies in important parts on theories and results from research literature relating to the target domain. This literature ranges from agent-based simulation specific works over cognitive and social psychology literature to innovation or social network research. We tried as much as possible of our theory driven assumptions to be supported by, first, the general *Household* agent architecture and its perception-action-loop, and second, a psychologically plausible cognitive decision making architecture of the agents.

The general perception-action loop of the *Household* agents follows the standard agent theory (Gilbert and Troitzsch 2005; Railsback and Grimm 2011). Each of the agents possesses an individual profile which they obtain during the initialisation step. In the first step of the perception-action loop, the sensor query step, each agent perceives its physical, social, and legal environment. This allows it to adapt to the current situation. In the options step, the agents pre-select that plan set which can be relied upon during the decision process. For a full-blown deliberate decision making, the subjective expected utilities for each plan are calculated in a subsequent filter step, before the actions associated with the one chosen plan are executed and the new values are exported to the partner models. Then, the cycle starts back again with a sensor query step.

Domestic water use has strong habitual components (i.e. much of daily drinking water use is not triggered by any conscious decision making), while there also are important deliberate decisions, e.g. when adopting water saving technological innovations or changing one's habits. The *Household* model thus provides representations for both of these processes: a bounded rationality based deliberate decision making mechanism on one hand and a habit component. While all *Household* agents have different attributes (like their milieu, household size, etc.), they share the same set of action options, and the decision making mechanisms. The route through decision making, however, depends on the agent attributes, thus resulting in a large variety of agent behaviour.

Habitual actions are represented by executing the agent's initialised standard behaviour, or, in case of extraordinary events (see below), the behaviour chosen in the last time step (Aarts and Dijksterhuis 2000). Both habits and reactions to extraordinary events are lifestyle-specific.

In the *Household* model, deliberative decision making is implemented by calculating the subjective expected utility (SEU) of all alternatives, which are plans in the set of known plans. To implement the decision process in a psychologically plausible manner, we draw on Ajzen's (1991) Theory of Planned Behaviour. It describes the formation of intentions that lead to actions with a certain probability by three components that are specific to each action: (a) The *attitudes* emerge from knowledge about the expected value of action consequences, (b) a person's *subjective norm* reflects the opinion of significant others and the person's willingness to comply with each of them, and (c) the *perceived behavioural control* describes the subjective perception of action barriers. Since this theory is action and situation specific, it has to be instantiated for every deliberative action in every agent anew. This reflects well the situated nature of decisions.

Most of the time, the *Household* agents behave according to their habits. Habits are formed through repeating the same action, originating in a deliberative process (Aarts and Dijksterhuis 2000). However, if important external events occur, agents will reconsider their behaviour. For instance, very high or very low temperatures cause many people to increase the frequency with which they shower or bath. Warnings about water shortages issued by water suppliers, the media, or local officials may cause an enhanced awareness to limiting water usage. Finally, agents rethink their water usage habits (and possibly, but by far not in all cases choose another way of behaving) if water prices show a considerable increase compared to the last time step.

Thus, if the agents' sensors signal the occurrence of special events like the ones just mentioned, full deliberative decision making is triggered. For example, the plan "shower frequency" becomes the target of a thorough decision process if the water price is raised by 5 % or more, if there is a drinking water quantity flag signalling water scarcity, or if the daily average temperature (day and night) rises above 10 °C. In our model, we call warnings about water shortages issued by water suppliers "water quantity flags". Such a quantity flag is supposed to come in four levels to mimic different psychological levels of communicating the severity of water scarcity, where level 1 means "no shortage", level 2 "news in print media or radio about a water shortage", level 3 "specific appeals from a commune official to save water", and level 4 finally "manifest water scarcity and supply by tank vehicles".

In the *Household* model, there exists a third decision mode: heuristic decision making (Gigerenzer et al. 2001). In its complexity, it is between full SEU decision making according to Ajzen (1991) and habitual actions. It only uses few pieces of information, in our model mostly stemming from the social network, thus mimicking a strong social comparison component. Heuristic decision making comes into play with the innovation adoption model described in the next paragraph. An extended and updated version of all decision modes mentioned here is available in the LARA architecture (Lightweight Architecture for Rational citizen Agents; Briegel et al., 2012) that is available from Sourceforge (at http://lara-framework.sourceforge.net).

The model distinguishes ten types of domestic water usage, from taking a shower, taking a bath or doing the laundry to flushing the toilet. This is part of the mission to produce an overall water use from individual use decisions. Moreover, this allows estimating the degree of pollution of waste water which is fed back from the households into the modelled water cycle. To estimate the water required by each of these usages, the efficiency values for different devices were obtained from the manifold literature on drinking water supply (e.g. Lehn et al. 1996). As important aspect of a changing overall water use, the installation of up-to-date, water saving household appliances can be identified. Thus, the model implements the decision making about the adoption of such devices as well. The agents decide upon competing water-use technologies in three areas: (a) shower heads (standard vs. water-saving vs. hydromassage shower that uses *more* water than the normal one), (b) toilet flushes (direct flush vs. standard tank vs. stop button vs. dual flush tank), and (c) rain harvesting systems (yes vs. no).

Shower heads are appliances which, from time to time, have to be replaced by newer, most probably more efficient ones, so 1 % of the agents decide every month about the acquisition of a new shower head. This value was considered a good estimate for the average innovation rate. Depending on technology and agent attitude towards progress and modernisation (i.e. its lifestyle), one of two decision algorithms is used to decide upon the technologies. While members of the Postmaterial and the Leading lifestyles are supposed to take deliberative decisions independent of the type of innovation, Traditional, Mainstream, and Hedonistic lifestyles are thought to make their choice dependent on the cost of the innovation: expensive rain water harvesting systems get full attention through deliberative decision making, while decisions about all other technologies are made in a heuristic way (Schwarz and Ernst 2009).

To take into account the spreading of innovations, agents are connected via an artificially generated social network (Schwarz and Ernst 2009). As the empirical surveys conducted for the agent-based model did not include an analysis of the subjects' social network, its generation was mostly inspired by the literature. As a growing number of social networks are found to have Small-World characteristics, the assumption was made that the social network of the agents has the characteristics of a Small-World network as well. The basic algorithm for generating the social network is the Small-World algorithm by Watts and Strogatz (1998) and was modified to include both spatial proximity and affinity to other lifestyles. The extensions made are threefold (see Schwarz and Ernst 2009): (1) Spatial proximity: the nearer two agents are, the more likely is a connection between these two agents, and the higher their social status, the more likely is a connection to an agent further away. (2) Social inclusion: agent types have a different number of peers within their closest social network (Postmaterialists and Social Leaders: 15, Hedonistic milieus: 10, Traditionals and Mainstream: 5). The rather low absolute number of agents within an individual social network is due to the fact that investment decisions may be based upon the opinion of rather few personal contacts (Fischer 1982). (3) Social Leadership: Social Leaders and—to a lesser extend—Postmaterialists and Hedonistic milieus are more likely to influence the decision of other agent types due to their role as opinion leaders within the social system.

To further test the influence of network structure on simulation results, a sensitivity analysis was carried out. We tested (1) several randomly generated versions of the extended network described in the text, (2) pure Small-World networks following the original algorithm, (3) randomly generated networks, and (4) a network based solely on spatial proximity. First of all, results indicate that network structure has some impact on the diffusion rate. Second, this impact is only relevant after innovations have spread to a certain extent (see Schwarz and Ernst 2008).

Besides providing the domestic water use and modelling the spreading of innovations in the DANUBIA context, the *Household* model is also responsible for simulating the perception and processing of water related risks (Seidl 2009; Seidl and Ernst 2008). Water related risks refer to either slow events like reduction of the groundwater levels or to highly visible sudden events like floods. Both receive very different media coverage. In addition to psychological work on risk perception dealing with these types of risk (Slovic 2000), this model drew on literature relating to the consideration of future consequences of one's actions (the CFC scale; Strathman et al. 1994), ideological preconceptions, psychological hygiene and suppression of uncomfortable thoughts, and psychological orientors (Bossel 2000).

The orientor concept derives agent motivations from their success in given environments (like resource scarcity, environmental variety, insecurity and the like) and the agents' subsequent adaptation processes. The risk perception module provides a perceptional sequence, where environmental events are first filtered by personal experience, perceived relevance of the event, and the consideration of future consequences. Then, their relation to the orientors is examined. Finally, postulated psycho-hygienic factors like coping, cognitive dissonance and defence mechanisms bring a further evaluation, before the agent gets to actively show some adaptation behaviour. The result shows that exceptional environmental events dampen an agent's well-being, but also that it recovers after some time. The resulting module cannot be reproduced in full detail here. Instead, the author has to refer to the description in Seidl (2009).

5.4.4 Visual Structuring Technique

To back up some of the theoretical surplus meaning in the risk perception module, visual structuring techniques (Scheele and Groeben 1988) were used to assess the more subtle aspects of risk related knowledge and its cognitive processing, as well as the motivational and emotional aspects of it.

5.4.5 A Lifestyle Typology

Lifestyles are a sociological concept (Bourdieu 1984) to categorize people or households according to their values and typical behavioural patterns, on top of their socio-demographic status. The lifestyle typology used here refers to the Sinus-Milieu® concept (Sinus Sociovision 2012), one of the leading lifestyle approaches in marketing in Europe. The Sinus-Milieus® were developed by the marketing company Sinus Sociovision. They divide the German population into ten so-called

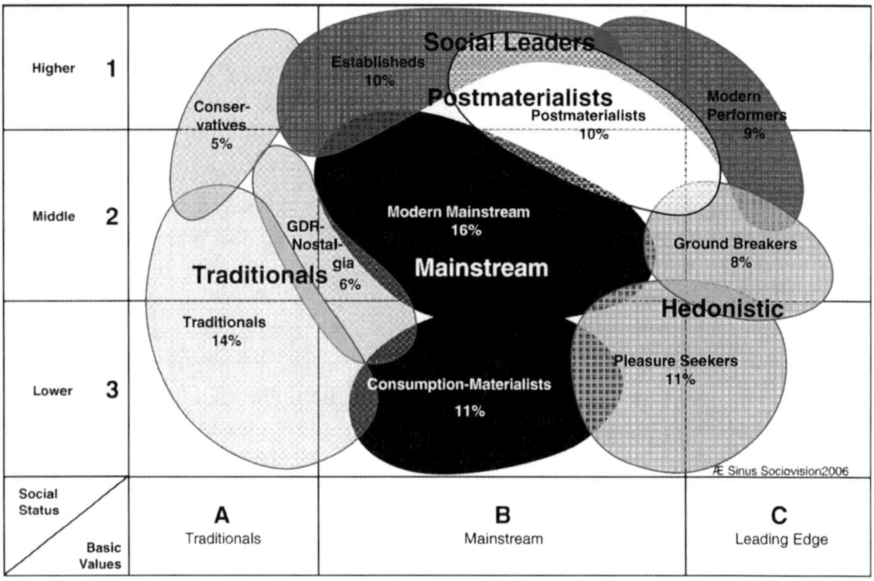

Figure 5.3 *The lifestyles and aggregated lifestyle groups according to the milieu classification of Sinus Sociovision.* The lifestyles spread along two dimensions: The classic social status found on the ordinate, and the basic values dimensions on the abscissa. Further explanation in the text. (Adapted from Sinus Sociovision 2006)

Sinus-Milieus® (Fig. 5.3). Apart from Germany, the Sinus-Milieus® are available for a large number of countries, many of them in Europe, but also including China or Canada. Though the concept of lifestyles is a static one, the Sinus-Milieus® are updated every couple of years, with the latest version stemming from 2011 for Germany. In the DANUBIA project, the classification of the year 2004 has been used throughout the project to maintain consistency.

These milieus, in the outset, have been constructed from extensive interviews on basic values and preferences, but also from the classification of their home furnishing and accessories as shown on photographs taken during the interviews. The dimension "basic values" adds considerably to the explanatory power of the concept and can be applied to a number of domains such as purchasing behaviour or innovation diffusion. To apply the Sinus-Milieus® in an empirical context, a set of 40 questions is provided which can be added to any survey or interview. These data are then analysed and coded into the ten lifestyles by the Sinus Sociovision company, with the exact algorithm remaining a business secret.

Establisheds are self-confident and think in terms of success and feasibility, while Modern Performers are the young and unconventional elite. Postmaterialists have liberal and postmaterial values, as well as intellectual interests. The old German educated class finds itself in the Sinus-Milieu® Conservatives with humanistic values and cultivated forms. Traditionals prefer security and orderliness, while GDR-Nostalgia believe in socialist visions of solidarity and justice. The modern mainstream aims at professional and social establishment and is very status-oriented, while consumption-materialists feel socially discriminated and aspire to the consumption patterns of the Mainstream. Experimentalists are very individualistic and see themselves as lifestyle avant-garde. Pleasure seekers have a low social status and refuse to accept the expectations of a performance-oriented society (Sinus Sociovision 2006).

Apart from having an empirically founded agent classification, the lifestyle typology used has the advantage of providing a number of scientific studies about environmental behaviour. Their results provide additional insights that can directly be linked to the agent types in the *Household* model. Among these studies count e.g. the bi-annual survey on environmental attitudes in Germany (Umweltbundesamt 2009) or a study about purchase of energy saving devices (Gröger et al. 2011).

5.4.6 Aggregation of Lifestyle Groups

For the present study, the original ten Sinus-Milieus® were clustered into five lifestyles groups: Postmaterialists, Social Leaders (encompassing Establisheds, Modern Performers), Traditionals (containing Traditionals, Conservatives, GDR-Nostalgia), Mainstream (with Modern Mainstream, Consumption-Materialists), Hedonistic (Pleasure Seekers, Ground Breakers). The lifestyle groups are proposed by the authors of the milieu classification themselves as one empirically sensible aggregation.

We used the reduction to five lifestyle groups to provide for a sound empirical foundation with data to inform the parameterisation of the agents in terms of attributes and behaviour (see next section). Agent types in the *Household* model thus represent—according to the empirical findings—one aggregated lifestyle group each.

5.4.7 Introducing Information About Agent Attributes (M2) and Agent Behaviour (M3)

5.4.8 Surveys

Several surveys (written questionnaires sent out by mail) with more than 1,500 respondents were carried through in the context of the project. As mentioned above, the gathered information about agent attributes and agent behaviour could directly be fed into the agent classes that had been defined by the lifestyle classification, since each survey also included the discriminative items to allow for the lifestyle attribution of each respondent.

The first survey specifically collected data about habitual behaviour: Water use (including its daily or weekly frequency), repairing of water appliances, choice of technology when replacing an appliance, individual cognitive and motivational aspects like attitudes towards saving water, water related conflicts, water related knowledge, and information about household members, appliances in the household, or household budget (Ernst et al. 2008). Based on this survey, for example, the shower length could be set to 6 min as a mean that had been suggested by our empirical data.

In a follow-up survey, aspects relating to the personal significance of water (the so-called water culture) were added and some other items clarified. Questions about the expected individual response to exceptional events (heat waves, water shortages, rising prices) were included as well. A more specific survey was also carried through to gather data about the consideration of future consequences, the awareness of climate change, responsibility, and psychological hygiene to support the implementation of the risk perception module of the *Household* model.

Another larger survey (Schwarz and Ernst 2006, 2009) was carried out to determine which criteria the actors take into account when deciding whether or not to buy water-saving devices for their homes, i.e. whether or not to adopt a certain technology. It was found e.g. that there are significant differences in the amount of information that lifestyle groups consider when making these decisions: the more modern and well-educated lifestyle groups are, the more their decisions rest upon a wider range of criteria. Moreover, these groups tend to take environmental aspects under consideration while the more conservative lifestyle groups are more sensitive to changing prices and the choices made by their peers. This survey allowed to directly empirically ground some of the relevant model parameters. This process covered two aspects: the assessment of innovation characteristics and the weights of decision factors. Some of the weights could directly be derived from the structural equation models computed to explain the empirical data (Schwarz and Ernst 2009).

5.4.9 Expert Workshops

By far not all of the model's parameters could be determined by the survey, though. There are different profiles for each of the implemented lifestyles. They differ with respect to the agents' perceived importance of the environment, of prices, and of the behaviour or opinion of peers in the acquaintance or family networks. Each of these is represented by a value in the agent profile. These values are inherited with the agents' lifestyle parameter and represent an important individual factor in the rational-choice decisions.

To get the best estimates for these values, we conducted two expert workshops with members of the development team of the Sinus-Milieus®. They considered the large array of partly unpublished studies dealing with lifestyles and environmental behaviour and drew analogies to the water use domain. The resulting estimates are values that mirror the lifestyles' characteristics relative to each other.

5.4.10 Telephone Interviews

A dozen telephone interviews with owners of rain harvesting systems (still quite uncommon in Germany) were conducted to learn more about the specific reasons of their installation. This was then used to abstract rules for the diffusion of more expensive water-related technologies (Schwarz and Ernst 2009)

5.4.11 *From Agent Types to the Artificial Population (Upscaling) (M5)*

As ingredients to upscaling, we have introduced an agent typology as well as extensive data relating to the domain of the model and the typology at the same time. What is still missing is the spatial distribution of the typology. Such data are exclusively provided by the combination of the Sinus-Milieus® with a spatial mapping of these lifestyles provided for Germany by the marketing company Microm® (Micromarketing Systeme und Consult GmbH, http://www.microm-online.de). These data are organised in so-called market cells that encompass up to 150 households (Microm 2012). In each market cell, the distributions of the Sinus lifestyle typology and the household sizes are known. These data are based on extensive geo-spatial research. The exact process, however, is not disclosed.

These data represent the spatial basis for the *Household* model. To use these data, the market cells had to be translated into the proxel (i.e. square kilometre) logic of the DANUBIA framework, by aggregating them based on their geo-referenced position. This was done in an initialisation process at the beginning of each model run. Every inhabited grid cell thus hosts five household agents, with each agent representing all households of that specific type. Differences between the lifestyles regarding certain parameters (e.g. income, environmental awareness) are represented in the agents' profiles. This initialisation thus builds the bridge needed to get an upscaled agent population that mimics the area's population not only in its spatial distribution, but within this distribution also in its characteristics and behavioural preferences.

The total water demand for one proxel is computed as the result of the individual water demands of each of the lifestyle type agents multiplied by the number of households of this type per proxel. This is based on data about the specific percentage of households of a certain lifestyle for each inhabited grid cell stemming from the geo-referenced data. Within each proxel, the distribution of household sizes for each agent type is also known and enters into the computation of water demand. This aggregated individual consumption defines the dynamically changing water demand on the proxel level and as such the micro-foundation of the macro-phenomenon to be modelled.

The modelled drinking water demand was compared against statistical data of domestic water consumption on the municipal level for the years 1998, 2001, 2004 and 2007 (Elbers et al. 2010). This validation was performed with the uncalibrated model against official statistical water use data on a monthly basis for a total of 1,415 municipalities that correspond to ca. 90 % of all inhabited proxels in the simulation area. The median of all individual differences over space and time between modelled and actual demand was 11 %, while the total absolute difference of the model and the total water consumption of the area was 1 %. This means that the model fit well the overall water consumption while the exact consumption per proxel resp. per municipality deviates more, probably due to local factors not considered in the model. To further improve the model fit against these data, the inclusion of e.g. qualitative factors like the existence of small businesses in some places and thus explaining a higher water use there would be preferred over just numerically calibrating the model.

5.5 Discussion: Lessons Learned

In this chapter, we have presented the *Household* model within the DANUBIA framework. This model is based on a number of combined data sources, ranging from extensive surveys to a spatially referenced lifestyle-based agent typology. It is meant to model domestic water use and water related risk perception in a spatially explicit way. The model is coupled at run time with the other models in the DANUBIA system, but can also be run stand-alone for the investigation of specific questions or for testing purposes.

We have described the main entities and state variables of the *Household* model together with the result presentation. The lifestyle typology played a major role in defining the agent types. Data to inform about agent attributes were collected from a broad range of sources. The gap between these data and a spatial representation of the typified population was bridged by commercially available data about the spatial distribution of lifestyles in space.

The combination of lifestyle differentiation and spatial explicitness provides some power to simulated scenarios. The *Household* model is driven by a sequence of scenario building blocks that can be freely combined. These scenario drivers encompass aspects from natural science as well as from social science. First, climate change projections from global climate models set the boundaries for the model runs. Then, one of a set of statistically selected realisations of regionalised climate model runs can be chosen. Furthermore, a number of projections of societal developments (societal scenarios) have been developed that influence the societal reactions to the environmental drivers. Finally, intervention scenarios represent more specific measures taken in the simulated run, e.g. policies, or technological advances driven by economic factors.

The scenario runs produced with the DANUBIA framework and the *Household* model can be analysed with regard to their spatial and temporal patterns. Figure 5.2 represents one example how movies can be used to show how the spatial evolution of some variable can be combined with graphs depicting the trend of its accumulated values over time. All in all, the approach of providing different climate and societal scenarios enables adequate analyses of the heterogeneous environment in a river catchment and complex human behaviour. The additional implementation of optional interventions, which can be edited and extended with ease, renders the tool more user friendly and thus allows for even more meaningful decision support for policy makers. Nevertheless, DANUBIA should not be mistaken as a planning tool. It rather indicates where, how and to what extent adjustments and interventions may become necessary (Soboll et al. 2011).

Unfortunately, the Sinus-Milieus® and their spatial differentiation are both provided by commercial marketing companies and therefore neither the questionnaire nor the rules to generate the typology are publicly available. Researchers have to pay for using this typology in their empirical studies. For the work described in this paper, the marketing company provided the questions to include them into the questionnaire on water-related innovations. The answers of all respondents relating to the Sinus-Milieus® were sent back and coded into lifestyles by this company. This procedure is on the one hand a clear disadvantage of using commercial marketing instruments for research purposes and their publication.

However, this approach is—to the author's knowledge—the only one readily available to produce a highly resolved spatially explicit model including a differentiated agent population. Moreover, there are numerous published (like the survey on environmental attitudes in Germany; Umweltbundesamt 2009) and unpublished studies building on the lifestyle classification used in the research presented in this chapter. This allows drawing analogies and comparing to other fields of research, though most of this work is certainly aimed at commercial marketing and not at environmental behaviour.

The application of the milieu approach employed here has the advantage of providing a credible and fairly robust typology that has been tested in many domains. There is one drawback: The robustness of such a general lifestyle typology is paid with a somewhat reduced explained variance for any specific, e.g. environmental behaviour like water use. Researchers have produced numerous context specific typologies, including some based on the lifestyle approach, for domains like mobility behaviour, energy use, and the like. These classifications are well fitted to the respective domains under investigation and thus maximise their predictive value, by design. However, each of these studies cannot be compared to any other just because of the individual typology. Moreover, the bridge to a spatial representation allowing for upscaling the data is missing in these specific typologies. All in all, this has led the author to the conclusion that using such a general lifestyle typology is an approach to spatially explicit agent based simulation worth being considered.

The *Household* model in the DANUBIA framework can be extended in several ways. Future developments could include a focus on policy options or on spatial communication patterns between citizens, e.g. to better investigate innovation processes relating to environmentally relevant technologies. This could—together with the feedback between natural and social model components—give new scientific insights as well as additional practical value of modelling social dynamics in a spatially explicit way.

Acknowledgements The author likes to thank the German Ministry for Education and Research (BMBF) for three consecutive grants to support GLOWA-Danube (Environmental Psychology) that helped to develop and refine the generic method presented here. We also acknowledge the support from Sinus Sociovision and Microm® Micromarketing Systeme und Consult GmbH for the research presented here. The author would also like to thank the numerous persons that helped carry through the research presented here, including Wolfram Mauser, Nina Schwarz, Silke Kuhn, Roman Seidl, Michael Elbers, Carsten Schulz, and Daniel Klemm.

References

Aarts H, Dijksterhuis A (2000) Habits as knowledge structures. Automaticity in goal-directed behavior. J Pers Soc Psychol 78(1):53–63

Ajzen I (1991) The theory of planned behavior. Organ Behav Hum Decis Process 50(2):179–211

Barth M, Hennicker R, Kraus A, Ludwig M (2004) DANUBIA: an integrative simulation system for global change research in the Upper Danube Basin. Cybern Syst 35(7–8):639–666

Barthel R, Janisch S, Schwarz N, Trifkovic A, Nickel D, Schulz C, Mauser W (2008) An integrated modelling framework for simulating regional-scale actor responses to global change in the water domain. Environ Model Softw 23:1095–1121

Bossel H (2000) Policy assessment and simulation of actor orientation for sustainable development. Ecolog Econ 35(3):337–355

Bourdieu P (1984) Distinction. A social critique of the judgment of taste. Harvard University Press, Cambridge, MA

Briegel R, Ernst A, Holzhauer S, Klemm, D, Krebs F, Martínez Piñánez A (2012) Social-ecological modelling with LARA: A psychologically well-founded lightweight agent architecture Seppelt R, Voinov A A, Lange S, Bankamp D (eds.) International Environmental Modelling and Software Society (iEMSs) 2012 International Congress on Environmental Modelling and Software. Managing Resources of a Limited Planet: Pathways and Visions under Uncertainty, Sixth Biennial Meeting, Leipzig, Germany. http://www.iemss.org/society/index.php/iemss-2012-proceedings.

Elbers M, Kuhn S, Klemm D, Ernst A (2010) Domestic water use under social and climatic change: projections of a coupled spatial simulation system. Proceedings of the 3rd world congress on social simulation, Kassel, Germany

Ernst A (2002) Modellierung der Trinkwassernutzung bei globalen Umweltveränderungen—erste Schritte. Umweltpsychologie 6(1):62–76

Ernst A (2009) Interaktion, Dynamik, Raum—Komplexe agentenbasierte Modelle in der Umweltpsychologie. Umweltpsychologie 13(1)84–98

Ernst A, Schulz C, Schwarz N, Janisch S (2005) Shallow and deep modelling of water use in a large, spatially explicit, coupled simulation system. In: Troitzsch K (ed) Representing social reality. Fölbach, Koblenz, pp 158–164

Ernst A, Schulz C, Schwarz N, Janisch S (2008) Modelling of water use decisions in a large, spatially explicit, coupled simulation system. In: Edmonds B, Hernández C, Troitzsch KG (eds) Social simulation technologies: advances and new discoveries. Idea Group Inc, Hershey, pp 138–149

Fischer C (1982) To dwell among friends: personal networks in town and city. The University of Chicago Press, Chicago

Gigerenzer G, Todd PM, the ABC Research Group (2001) Simple heuristics that make us smart. Oxford University Press, Oxford

Gilbert N, Troitzsch KG (2005) Simulation for the social scientist. Open University Press

Gröger M, Schmid V, Bruckner T (2011) Lifestyles and their impact on energy-related investment decisions. Low Carbon Econ 2:107–114

Hennicker R, Ludwig M (2006) Design and implementation of a coordination model for distributed simulations. In: Tagungsband Modellierung 2006—Forschung & Praxis, 7. Fachtagung des Querschnittfachausschusses Modellierung in der Gesellschaft für Informatik eV (GI), Lecture Notes in Informatics, Vol. P-82, Series of the Gesellschaft für Informatik (GI). pp 83–97

Kneer JErnstA, Eisentraut R, Nethe M, Mauser W (2003) Interdisziplinäre Modellbildung: Das Beispiel GLOWA-Danube. Umweltpsychologie 7(2):54–70

Lehn H, Steiner M, Mohr H (1996) Wasser—die elementare Ressource: Leitlinien einer nachhaltigen Nutzung. Veröffentlichungen der Akademie für Technikfolgenabschätzung in Baden-Württemberg. Springer, Berlin

Ludwig R, Mauser W, Niemeyer S, Colgan A, Stolz R, Escher-Vetter H, Kuhn M, Reichstein M, Tenhunen J, Kraus A, Ludwig M, Barth M, Hennicker R (2003) Web-based modelling of energy, water and matter fluxes to support decision making in mesoscale catchments—the integrative perspective of GLOWA-Danube. Phys Chem Earth 28:621–634

Mauser W (2000) GLOWA-DANUBE—Integrative techniques, scenarios and strategies regarding global changes of the water cycle. Case study: Upper Danube catchment area. http://www.glowa-danube.de/publikationen/umbrellas/umbrella_2000.pdf. Accessed 20 July 2012

Microm (2012) Microm consumer marketing. http://www.microm-online.de/Deutsch/Microm/index.jsp. Accessed 20 July 2012

Railsback SF, Grimm V (2011) Agent-based and individual-based modeling: a practical introduction. Princeton University Press, Princeton, NJ

Scheele B, Groeben N (1988) Dialog-Konsens-Methoden zur Rekonstruktion subjektiver Theorien. Francke, Tübingen

Schwarz N, Ernst A (2006) Using empirical data to build an agent-based model of innovation diffusion. In: Proceedings of the workshop on agent-based models of market dynamics and consumer behaviour, Surrey, GB, 17–18 Jan 2006

Schwarz N, Ernst A (2008) Die Adoption von technischen Umweltinnovationen: Das Beispiel Trinkwasser. Umweltpsychologie 12(1):28–48

Schwarz N, Ernst A (2009) Agent-based modeling of the diffusion of environmental innovations—an empirical approach. Technol Forecast Soc Change 76(4):497–511

Seidl R (2009) Eine Multi-Agentensimulation der Wahrnehmung wasserbezogener Klimarisiken. Metropolis, Marburg

Seidl R, Ernst A (2008) Perception of climate change risks: a multi-agent simulation. J Psychol 43(3–4):249–249

Sinus Sociovision (2006) Informationen zu den Sinus-Milieus 2004

Sinus Sociovision (2012) Sinus-Milieus. http://www.sinus-institut.de/en/. Accessed 20 July 2012

Slovic P (ed) (2000) The perception of risk. Earthscan Publications, London

Soboll A, Elbers M, Barthel R, Schmude J, Ernst A, Ziller R (2011) Integrated regional modelling and scenario development to evaluate future water demand under global change conditions. Mitig Adapt Strat Glob Change 16:477–498

Strathman A, Gleicher F, Boninger DS, Edwards CS (1994) The consideration of future consequences—weighing immediate and distant outcomes of behavior. J Pers Soc Psychol 66(4):742–752

Umweltbundesamt (2009) Umweltbewusstsein und Umweltverhalten der sozialen Milieus in Deutschland. Dessau-Roßlau

Watts D, Strogatz S (1998) Collective dynamics of 'small world' networks. Nature 393:440–442

Chapter 6
Parameterisation of AgriPoliS: A Model of Agricultural Structural Change

Christoph Sahrbacher, Amanda Sahrbacher and Alfons Balmann

6.1 Model Description[1]

This model description follows the ODD protocol (Overview, Design concepts and Details) as proposed and described in Grimm et al. (2006) and Grimm et al. (2010). The ODD protocol allows a standardized description of agent-based models what improves the clarity and the comparability of models. The following ODD protocol has already been published by Sahrbacher et al. (2012).

6.1.1 Overview

Purpose The main purpose of the model is to understand how farm structures change within a region, particularly in response to different policies. Structural change in agriculture is the result of farms' individual decisions. Each farm decides what to produce and how much, whether it is worth renting additional land or rather releasing some of it, or even to exit agriculture if necessary. Depending on their individual situation and their neighbourhood, farms react differently to the same

[1] AgriPoliS has been initially developed by Happe, Balmann and Kellermann based on Balmann (1997) and has been first published in Happe (2004) and Happe et al. (2006). This description of AgriPoliS following the ODD-protocol is based on Kellermann et al. (2008) the last update of the model description.

C. Sahrbacher (✉) · A. Sahrbacher · A. Balmann
Leibniz Institute of Agricultural Development in Central and Eastern Europe, Halle, Germany
e-mail: sahrbacher@iamo.de

A. Sahrbacher
e-mail: a.sahrbacher@iamo.de

A. Balmann
e-mail: balmann@iamo.de

A. Smajgl, O. Barreteau (eds.), *Empirical Agent-Based Modelling – Challenges and Solutions*, DOI 10.1007/978-1-4614-6134-0_6,

environmental conditions or policy changes. The heterogeneity in the farm population allows covering these individual responses. Land markets play a crucial role in agricultural structural change. Because the dynamics on land markets are mainly determined by the interactions between individual farms, an agent-based approach offers many advantages for creating an explicit land market.

Entities, State Variables and Scales The model comprises different hierarchical levels: farms, farm population, plots, landscape, markets and political environment. The model's main entities are farms, production activities, investment objects, plots,[2] markets and the political and economic environment.
Farms are income or profit maximising entities and characterised by state variables such as an ID-number, age, managerial ability, location in space and the amount of production factors (land, capital, labour) they own, or rent. A farmer actively manages a farm during 25 years (maximum duration of management) and hands his farm over to a possible successor after this period, i.e. AgriPoliS considers generation change. Age corresponds to the number of years a farmer is already managing his farm. Whereas managerial ability, location and the amount of family labour stays constant over time, the amount of land and capital can change. Capital is further separated into liquid assets, fixed assets (investment objects) and entitlements for production (milk quota). Farms can choose out of a list of production activities and investment options in order to make an optimal use of their production factors during the production process.

Regarding production activities, it can be further differentiated between plant production activities, livestock production activities and auxiliary activities. All production activities are continuous decision variables and characterised by an ID-number, a revenue, production costs and a price flexibility. Production activities are associated to a production branch like in EU's Farm Accountancy Data Network (e.g. field crop, milk, grazing livestock and pig and poultry). This allows the classification of farms according to their production branch. Furthermore, production activities are associated to a type of investment which is necessary for the production. Labour requirement and the premium associated constitute state variables common to plant and livestock production activities. In addition to this, plant production activities require land of specific soil type and machinery as well as restrictions for crop rotation. Some plant activities can supply fodder for livestock. Livestock production activities require stable places and fodder. Furthermore, they are characterised by a livestock unit. Auxiliary activities are defined to allow an optimal use of capital and labour. Thereby, either overcapacities of production factors can be reduced (e.g. saving money, working off-farm, leasing quota) or scarce capacities can be borrowed, hired or leased in a short-term perspective (money, labour, quota, machinery). Only revenues and premium can change over time due to changes in the political and economic environment. Production costs vary among farms according to their managerial ability.

[2] What we name "plots" are individual equally-sized cells in the artificial region. All plots joined together constitute the artificial region, like fields constitute specific landscapes in the reality.

All investment objects are integer variables, i.e., they are not dividable and they are financed with a specific share of long-term borrowed capital. The interest rate of long-term borrowed capital is the same one for all investment objects. Further state variables of investment objects are: an ID-number, type of investment (e.g. hog house, cowshed, machinery, hiring a worker for a year or working off-farm for a year), production capacity, investment costs, maximum useful life, age, maintenance costs, labour saving due to size effects and technical progress associated with the investment. Technical progress depends on investment size.

Plots are basic elements of the grid representing an artificial landscape. The state variables of these basic land units are size, location, soil type, ownership, rental price, rental contract duration and age of the contract. Location, size and soil type of a plot are constant during the simulations whereas the ownership status, the rental price and the age of the contract can change over time. The soil type defines land of a specific quality (e.g. high, medium, low, etc.). Depending on the quality, land can only be used for specific production activities and generate a particular yield. Thereby, it is indirectly differentiated between arable land, grassland and non-agricultural land (settlements, roads, rivers, lakes and forests that cannot be used for agriculture). Plots are also characterised by their ownership status; they can be owned, rented or abandoned. Rental prices can change because an endogenous rental market is modelled. Rental contracts have a fixed duration[3] and are terminated when the age of the contract equals the duration. The contract duration is randomly drawn from a uniform distribution whereby a minimum and maximum contract duration is set. Taken altogether, the location of the plots, their size and the soil type form an artificial landscape.

The landscape is represented by five different layers: ownership, soil distribution, block, allocation and usage/field layers. The ownership layer defines the ownership of a specific plot. The second layer represents the distribution of different soil types across the landscape. The block layer replicates the distribution of contiguous areas of given soil types (e.g. arable land and grassland) thus reflecting the geophysical conditions in a region. Such a block is defined as a group of contiguous plots[4] of the same soil type that is separated from other plots of the same type by another soil type. The allocation layer represents how blocks are allocated to farms. Contiguous plots belonging to a farm are called sub-blocks. Plots in a sub-block can either be rented or owned by farms. The fifth layer reflects farms' land use, i.e. a field comprising contiguous plots used by a particular farm for a particular purpose (e.g. wheat production).

Markets are distinguished according to their scale and the applied rules. Regarding the scale, markets are either national and EU-wide, regional or regional and spatially organized at the same time. National and EU-wide markets both follow the

[3] There exists another type of rental contracts in AgriPoliS, namely plots can be "renegotiated" at the end of each production period. At the end of a production period for each rented plot, each farm decides either to keep the plot or to release it. The decision rule itself is based on the expected revenue of a plot in the next production period (Kellermann et al. 2008).

[4] Two plots are contiguous if they are adjacent to each other.

same rules which are applied to products like crops, oil seeds, sugar, capital, labour and services.[5] Products traded on regional markets differ from region to region and depend on the specific characteristic of the region modelled. Therefore, even the market rules can differ among products. At the regional scale, these are, e.g. a market for manure (Germany, France), a market for calves (Sweden, Västerbotten and Jönköping), or a market for milk quota (almost all regions). For regional spatially-organized markets it is assumed that goods are not traded on spot markets. This applies to the land market which is organised as an auction.

The political and economic environment mainly affects prices and payments for production activities over time. Price changes may be caused by changes in the agricultural policy or in the economy. In addition, there might be new obligations to be fulfilled so that farmers receive payments.

So far 22 different agricultural regions have been modelled with AgriPoliS.[6] Their size varies from 20,000 to 1.7 million ha. The number of farms varies between 511 and 45,000 farms. However, in general only a proportion of a region is simulated because of computational time constraints. The plot size is chosen based on regional and computational criteria and varies between 0.5 and 5 ha. For computational reasons the plot sizes should be as large as possible because the number of plots depending on the plot size and the region size is a crucial figure. The more plots there are, the more time is necessary for the allocation of free plots and therefore the longer a simulation will last. A simulation normally runs over 25 time steps (periods) where each time step is equivalent to one year.

Process Overview and Scheduling In each period the farms go through the following processes in the presented order: land auction, investment, production, update product markets, farm accounting, set policy, disinvest, exit decision, period results. The schedule is the same for each period. Each process begins when all farms are finished with the previous one. Farms' order stays the same in each single process. For land auction, investment, production and exit decision farms apply a Mixed Integer Programming model (MIP model, cf. Hazell and Norton 1986) in order to make their decisions in each step. The MIP model brings together farm's endowment of production factors, the production activities and the investment options.

In the first period all plots are distributed to farmers, thus land auction is not necessary. In the following periods free land is allocated among farmers via a sequential first-price auction. In each sequence, bids are made for only one plot, i.e., every farm selects the plot which is most valuable to it and then calculates the bid accordingly. The farm with the highest bid gets the plot it wishes. The auction is repeated afterwards until all plots are allocated or until there are no further positive bids. The fact that farms bid for the most valuable plot instead of all farms bidding at the same time for one plot avoids first-mover advantage. To consider complementarities bet-

[5] As the study regions are predominantly small, the effect on prices of nationally and internationally traded products should be marginal. Thus, the national and EU-wide markets are in almost all cases not used.

[6] An overview about six of these regions can be found in Sahrbacher (2011).

ween different soil types, the auction alternates between them. When a plot is rented a contract duration is assigned to it. The contract duration is binding, that is neither the land owner nor the land manager can terminate or renegotiate the rental contract during the entire contractual period. Based on their actual amount of land, farms calculate one after another whether they want to invest or not. After investment decisions are made, the available hours of labour, liquidity as well as stable places and machinery endowments are updated. In the production process each farm chooses the amount and combination of production activities in order to optimally use its production factors. The total amount of products is summed up and transferred to the market, where product prices are updated accordingly. The actual prices are then considered in farms' accountancies. For the farm accounting financial indicators such as income, profit, equity capital change, depreciation of buildings, withdrawals for unpaid family labour etc. are calculated. Unpaid family members withdraw a certain amount of money from the farm's profit for paying taxes and consumption. Remaining money is saved and increases equity capital. If withdrawals are higher than profit, equity capital is reduced. Accountancy data for each individual farm are written in output files and apart from this, data is aggregated for all farms as well. At the end of each period, farms receive information about the political and economic environment for the next year. This affects the farms' expectations for the future. In the disinvest process, the amount of production factors is updated, that is plots with terminated rental contracts are released and investment objects at the end of their useful lifetime are subtracted from the list of farm's production factors. Considering the updated production factors and their expectations regarding future policy and economic environment, farms decide whether to continue farming or not based on the objective to maximise profit or income. Therefore, they calculate their expected income from agriculture for the next year and compare it with the opportunity costs of their production factors. If the age of a farmer is equal to the maximum duration of management a generation change takes place. Therewith, opportunity costs of a possible successor are assumed to be 25 % higher because education and training are considered as specific and irreversible investments. If opportunity costs are higher than the expected agricultural income, farms exit the sector. They also do so if they are illiquid. Land released by terminated contracts and exiting farms is available for renting in the next period. The simulation terminates when the number of specified time steps is reached or if all farms stopped farming.

6.1.2 Design Concepts

Basic Principles The following economic concepts are considered in the model: profit or income maximisation, sunk costs, path dependency, economies of size, myopic behaviour, shadow price, transport costs, and opportunity costs.

According to the theory of *sunk costs* farms do not consider costs of assets (stables and machinery) because these costs arise anyway. This is a realistic assumption because a market for agricultural buildings does not really exist. Sunk costs affect farmers' behaviour and lead to path dependency as shown by Balmann et al. (1996). *Path dependency* describes a situation where a system is locked in and can only be abandoned at extremely high costs (Arthur 1989). *Economies of size* are implemented in the model so that investment costs per unit decrease with the size of the investment. Moreover, labour is assumed to be used more effectively with increasing size.

Farm decision making is *myopic* or *boundedly* rational (Simon 1955, 1956, 1996), that is, agents make decisions based on the information available to them, which can possibly even be wrong. The decision problem of the modelled farms is highly simplified compared to that of real farmers in that strategic aspects are not included. Indeed, individual farms in AgriPoliS have information on rents as well as product and input prices but they do not know about other farms' production decisions, factor endowments, size, etc. Farm agents are also boundedly rational with respect to expectations. In reality in a majority of cases, farm agents follow adaptive expectations. Merely policy changes are anticipated one period in advance and included into the decision making process.

To rent land, farms formulate a bid about how much they are willing to pay for renting an additional plot. This bid is based on the farm's *shadow price* of land. The shadow price is the value by which the farm income would increase if the amount of scarce production factors, here land is increased by one unit (cf. Hazell and Norton 1986).

Compared to regional models like RAUMIS and farm group models like FARMIS, AROPAj and FAMOS[7] which are mainly used to predict agricultural production, AgriPoliS considers space explicitly. Even if, the landscape in AgriPoliS is rather a model of the real landscape than an exact replication of it (e.g. GIS-based) this allows considering *transport costs* from the farmstead to the plots and thereby, spatial competition on the local land market.

Emergence Structural change and land price evolution emerge based on the individual decisions of all farms. How a farm develops can only be predicted to a certain extend based on the initialization (e.g. product prices and costs). A farm's development always depends on decisions of neighbouring farms.

Adaption Farms adapt to changing conditions on markets, their local environment and to policy changes by changing the combination of their production activities or production factors in particular by investing in new objects. Furthermore, they can grow by renting additional plots or shrink. Shrinking is not an active process. Farms release land only when a rental contract for a plot is terminated or if the farm is closed down. Surviving farms can try to lease the plot again or other plots by formulating a bid but it is not sure that this will be the highest bid and that they get the acceptance.[8] A final and irreversible adaption is the exit from agriculture.

[7] Henrichsmeyer et al. (1996), Jacobs (1998), Jayet et al. (2007) and Hofreither et al. (2005).

[8] There exists also another version of the land market where the rental contract length is infinite and land is only released by farmers if it is not profitable for them to use it. In this version of the land market farms can actively shrink.

Objectives Farms maximise their household income by using mathematical programming. Therefore, a one-period Mixed Integer Programming model is built including continuous production activities (hire employees on an hourly basis, grow crops on a certain proportion of farmland, borrow credits, etc.), integer investment options (build one stable and not parts of it, hire employees on a yearly-basis and not parts of them), farm's production factors and general production restrictions (cf. Hazell and Norton 1986).

Prediction Farms expectations about future prices are adaptive.

Sensing Farms are assumed to know their own state and the state of their investment objects and plots. They also get information about policy changes and price changes caused by policy changes one period in advance. Furthermore, farms sense the state of all plots in the region, and hence can determine which additional plot they wish to rent. However, they do not have any information about their neighbours.

Interaction In AgriPoliS, farms interact indirectly via the land market as well as via markets for products. On the land market, farms are directly competing with each other. Product markets can be influenced by the individual farmer's production decision as the supply of a product affecting its price. National and EU-wide product markets are coordinated via a simple price function with an exogenously given price elasticity. For regional markets other rules exist.

Observation AgriPoliS produces results at the sector level as well as for each individual farm at each time step. It delivers results related to economic indicators, production, and investment. Depending on the research question different indicators are analysed. Economic indicators such as profit per farm, economic land rent, rental prices or farm growth are systematically analysed. Results on production and investments are analysed to understand the economic effects.

6.1.3 Details

Initialisation The agricultural structure of a region is defined by the number of farms, livestock, total amount of land of different qualities, number of farms in specific size classes etc. Data about these characteristics are taken from regional statistics. To create a virtual region and a virtual farm population, a set of farms from the European Union's Farm Accountancy Data Network (FADN) in a reference year is simultaneously selected and up-scaled so that farms of this set represent the agricultural structure of a study region best. Thereby, the squared deviation between the sum of up-scaled farm characteristics such as farm size, number of animals etc. and the corresponding regional characteristics is minimised applying a quadratic programming algorithm (for details see Sect. 6.3.1). During the initialisation each selected farm is cloned according to its up-scaling factor. The cloned farms are then individualised with respect to managerial ability, location, age, age of assets, and the duration of each plot's rental contract. Values of these state variables are randomly

assigned within a range defined for each study region or simulation. Managerial ability is assumed to influence production costs; therefore costs vary within a predefined range differentiating farmers regarding their management performances. The age corresponds to the number of years a farmer is already managing his farm. It is assumed that a farmer manages its farm for 25 years and retires afterwards. The age of assets varies between zero and the useful lifetime. Regarding rental contracts, a minimum and a maximum duration are defined.
Data about production activities and investment objects also vary amongst study regions. They are taken from farm management pocket books. In order to fit the selected farms' production levels they have to be calibrated before AgriPoliS' initialisation stage. Details about the initialisation of the state values are documented together with the respective case studies (e.g. Sahrbacher 2011).

Input Changes in the political and economic environment are considered during a simulation run. Political changes, i.e. changes in the direct payment level or new regulations (modulation, capping, set aside obligation, livestock density etc.) are introduced based on upcoming policy reforms or hypothetical scenarios can be defined by the user. Data about product price changes are provided either by other models like the partial equilibrium model ESIM (European Simulation Model), or they are based on external scenarios or they are varied randomly to simulate price volatility. Political and economic settings vary from study region to study region and depend on the research question. They are documented with each application of the model.

Submodels Following the ODD protocol, this section aims at providing a full model description, especially in the form of mathematical equations and rules in order for the reader to fully understand the what's and how's constituting the inside of the model. All this material can be found in Kellermann et al. (2008).

6.2 Parameterisation Overview

The empirical application of this model follows the sequence outlined for case 3. Purpose of the model is to understand how farm structures change within a region. Structural change in agriculture is the result of farms' individual decisions. Thus, only one agent class was identified (i.e., farmers) based on expert knowledge (M1). Farms are not differentiated into types according to their behaviour because of the common assumption in agricultural economics (M3) that farmers maximize their profit. Indeed, other behavioural rules are plausible, for example utility maximization, a safety first model (Chavas et al. 2010) or maximization of labour input in the case of cooperative farms but no information is available about the distribution of those different behaviours. However, farms can be differentiated according to their attributes such as size of utilized agricultural area, product specific specialisation (e.g. field crops, dairy cows, grazing livestock, pig and poultry etc.), number of different livestock etc. Data about farms' attributes are from a survey that is annually conducted by the FADN (M2). It is decided based on census data characterising the study region and on expert knowledge which attributes are the most important to represent the agricultural structure of the study region. The number of observations in this survey varies depending on the size and structure of the study region. Within the 22 modelled regions the smallest number of observations was 35 and the highest 605. Indeed, this survey is representative at the country level but not necessarily at the regional level. Thus, the relevant agent types have to be selected from the survey data and up-scaled using regional census data. To identify the relevant agent types based on their attributes and to scale them up a specific method is used. This method automatically selects and weights those farms from the survey with which important regional structural characteristics can be represented best. That is "clustering" (M4) and up-scaling (M5) is done in one step. However, this procedure does not guarantee that regional crop production is correctly represented. Therefore, production activities and investment options specific for the study region have to be defined in a second step. Suitable data sources are standard farm management norms as provided, for example, by farm management pocket books in Germany e.g. from the Kuratorium für Technik und Bauwesen in der Landwirtschaft (KTBL) and others.

6.3 Technical Details

As mentioned, parameterization is done in two steps. First, the initial farm population is created. Then, further characteristics of the farm agents namely production activities and investment options are defined to represent the farms' production structure using mathematical programming. Therefore, the coefficients of production activities and investment options have to be calibrated to represent the farms' production which represents in total the production structure of the study region. In the following both steps are described in detail.

6.3.1 Selection and Up-Scaling of Farms to Represent the Regional Agricultural Structure

To create the initial virtual farm structure of a study region the following approach created by Balmann et al. (1998) and further developed by Kleingarn (2002) and Sahrbacher and Happe (2008) is applied. This particular approach requires two kinds of data: *first*, data about region's *general characteristics* such as total number of farms, total utilised agricultural land and total number of different livestock and its *structural characteristics* like number of farms per farm type or legal form, share of different soil types (e.g. arable and land grassland), number of farms in different size classes and number of livestock per herd size class. Potential data sources for aggregated regional data are statistical offices. *Second*, corresponding data about individual farms in the study region are needed. The FADN and the Integrated Administration and Control System (IACS) collect data about individual farms within the EU. The FADN additionally delivers economic indicators for each farm. Depending on the region's size, the number of observation from FADN can be about 100 or more farms. IACS delivers data about all farms in a study region but it is difficult to get access to this database and no financial indicators are provided. Even though data about all farms of a study region should be available it is better to select some of them because all farms are represented by the same mixed integer programming model which has to be calibrated that it reflects all modelled farms production. This means coefficients in this model have to be calibrated to represent all farms' production, which becomes more difficult with an increasing number of farms. An appropriate number of selected farms for modelling is 20–30.

Following Balmann et al. (1998), the up-scaling procedure can be explained as follows: *First*, the set of farms coming from FADN or other sources is put into a matrix. *Second*, the census data about the regional characteristics are added and defined as goal criteria. *Third*, an optimisation problem is formulated, which assigns weights to each farm and minimises the quadratic deviation between the sum of weighted farm characteristics and the respective regional characteristics. If some characteristics are more important than others, they can be prioritised. Negative weights are ruled out.

In mathematical terms, this procedure can be explained as follows (cf. Balmann et al. 1998):

Let $\mathbf{b} \in \Re^m$ be the vector of weights for m farms and let $\mathbf{y} \in \Re^n$ be the vector of n statistical goal criteria in the region. Furthermore, let $v_{i,j}$ be the contribution j of farm i, and $V \in \Re^{m*n}$ the matrix of contributions of all farms. From this, the vector of all goal criteria $\hat{\mathbf{y}}$ for the virtual region can be derived

$$\hat{\mathbf{y}}^0 = \mathbf{b}^T \mathbf{V}^0 .$$

Now a normalised matrix $\mathbf{X} \in \Re^{m*n}$ can be constructed with

$$\mathbf{X} = \left[a_j \frac{v_{i,j}}{y_j} \right]_{\substack{i=1,\ldots,m \\ j=1,\ldots,n}}$$

and a_j as the priority level of criterion j in a region, or $\mathbf{a} \in \Re^m$ as the vector of weights of all criteria in a region. The vector of weights **b** then results from the minimisation problem

$$\min_{\mathrm{b}} \left\{ (\mathbf{Xb} - \mathbf{a})^T (\mathbf{Xb} - \mathbf{a}) \right\} \text{ with } \mathbf{b} \geq \mathbf{0}.$$

This problem can be solved with a quadratic programming algorithm.

	A	B	C	D	E	F	G	H	I	J	K	L	M
1				General characteristics						Structural characteristics			
2										Number of farms per legal form			UAA per legal form
3		Min weighting	Weighting of farms	Number of farms	Total UAA (ha)	Bulls	Dairy cows	Suckler cows	Breeding sows	Legal entities	Partner-ships	Individual farms	Legal entities
4	**Regional data**			**1852**	**479100**	**7451**	**96272**	**10500**	**26574**	**234**	**208**	**761**	**234**
5	Farm data	0	72	1	560	0	0	0	0	1	0	0	560
6		0	47	1	1195	0	0	0	0	1	0	0	1195
7		0	115	1	20	0	0	15	0	0	1	0	0
8		0	53	1	380	0	140	0	0	0	1	0	0
9		0	46	1	110	0	0	0	15	0	0	1	0
10		0	44	1	150	30	95	0	0	0	0	1	0
11													
12	**Sum of weighted farm characteristics** =Sumproduct($C5:$C10,D4:D10)			**377**	**119**	**1320**	**11600**	**1725**	**690**	**168**	**90**	**130585**	**96485**
13	Weighted squared relative deviation =(D12/D4-1)^2*D14			1.269	0.483	1.354	1.547	1.397	1.897	0.074	1.555	1.058	0.378
14	Weighting for prioritisation			2	2	2	2	2	2	2	2	2	2
15	**Relative deviation to regional data** =D12/D4-1			**-80%**	**-49%**	**-82%**	**-88%**	**-84%**	**-97%**	**-19%**	**-88%**	**-73%**	**-43%**
16	max. relative deviation			100%	100%	100%	100%	100%	100%	100%	100%	100%	100%
17													
18	**Objective** =(sum(D13:M13)/sum(D14:M14))^0.5			**0.851**									

Source: Own description.

Figure 6.1 Exemplary scheme of the up-scaling procedure in MS Excel. (Source: Own description)

Technically this approach can be implemented in MS Excel by using the MS Excel Solver. Figure 6.1 shows the matrix with regional data about general and structural characteristics and the corresponding individual farm data. The list of characteristics is not comprehensive in this example. It can be extended by further characteristics described above. For further examples see also Sahrbacher (2011). For creating the matrix farm data has to be adjusted. The farm size has to be completely divisible by the plot size and depending on the region small livestock herds are deleted, e.g. if a farm has less than five animals per type. When the matrix is finished the mathematical programming problem can be solved to create the virtual region. The value of the virtual region's characteristics ("sum of weighted farm characteristics") is determined by weighting the farm's contribution to a characteristic and summing them up. The objective of the mathematical programming problem is to choose the weighting of the farms (b) such that the deviation of the virtual region's characteristics from the real characteristics is minimised. Therefore, cell D19 has to be set as objective in the Solver. The "variable cells" in which the weighting factors are written by the Solver are C5 to C10. Furthermore, two constraints have to be fulfilled. First, the weighting should be non-negative that is the values in cell C5 to C10 should be equal or bigger than zero (cells B5 to B10). Second, the maximum relative deviation of the sum of weighted characteristics from the regional data (cells D15 to M15) should not be higher than 100% (cells D16 to M16). Both constraints have to be set in the Solver. With a large number of farms the solver will find a solution after a while. Restarting the Solver several times will improve the solution. Furthermore, one can manually check whether there are similar farms. To reduce the number of farms similar ones can be deleted. Rerunning the Solver will give almost the same result.

6.3.2 Representation of the Selected Farms

To represent the behaviour and organisation of the selected farms a MIP model is built. The MIP model fulfils two tasks. The first task is to represent and reproduce the selected farms' observed production structure.[9] The second task is to provide options for alternative farm organisations, i.e. different investment options, working off-farm, hiring additional labour, contracting, and savings in order to maximize incomes or profits and to fulfil production constraints. Therefore, production activities, financing activities, auxiliary activities, investment possibilities, farm factor endowments, and restrictions to farming activities are all grouped in a matrix. It is assumed that farms maximise their household income. Figure 6.2 shows an exemplary matrix of the optimisation problem.

---

[9] Compared to highly-differentiated and detailed farm-based linear programming models, the optimisation model in AgriPoliS is aggregated. Yet, with respect to the objective of AgriPoliS, it is not the specific farming system which is of interest in this study, but rather a basic representation of central organisational characteristics, as well as financial and economic considerations.

Mixed-integer programme		Short-term loans/savings	Buy/sell variable labour	Hire contractor	Plant production	Livestock production	Set-aside land	Buy/sell manure	Buy/sell milk quota	Investment activities	Buy/sell fixed labour
		c	c	c	c	c	c	c	c	i	i
	Objective function	*Gross margin*									
Factor capacities	Liquidity (€)	x		x	x	x	x			x	x
	Min. equity capital reserve (€)				x	x	x			x	x
	Labour (h)		x		x	x	x	x		x	x
	Utilised agricultural area (ha)				x						
	Milk quota (litres)					x			x		
	Livestock capacities (places)					x				x	
	Machinery (ha)			x	x		x			x	
Other restrictions	Organic N-balance (kg N/ha)				x	x		x			
	Rape seed max. (% of UAA)				x		x				
	Sugar beet max. (% of UAA)				x						
	Set aside (% of UAA)				x		x				
	Fodder (ha)				x	x					
	Direct payments (€)				x	x	x				
	Stocking density (LU/ha)				x	x	x				

Notes: c = continuous activities, i = integer activities.

Source: Based on HAPPE (2004).

Figure 6.2 Exemplary scheme of a mixed-integer programme matrix. *c* continuous activities, *i* integer activities. (Source: Based on Happe 2004)

To represent the selected farms by means of a MIP model the following steps have to be carried out:

- Definition of regional production activities and their restrictions.
- Definition of regional investment options.
- Assignment of investment options to selected farms based on their size and the amount of their livestock husbandry.
- Identification of alternative production activities.
- Construction and compilation of the MIP matrix based on farms' specific factor endowments.
- Calibration of MIP model's parameters regarding the following criteria:
  - No new investments to occur in the initial setting, otherwise large deviation between observed and optimised production in the initialisation period.
  - Full use of farms' factor endowments (land, machinery and stables).
  - Limitation of losses, as farms would exit too quickly in AgriPoliS for illiquidity reasons.

Depending on the study region, production conditions differ and hence farmers are producing different crops or keep different kinds of animals. Furthermore, different stable types and sizes might be common in a study region. To identify typical production activities and investment options regional data sources, as well as individual farm data from FADN can be used. In addition, regional experts can be asked to verify the choice of typical production activities in the respective region. In order to properly represent the selected farms, production activities and investment options have to be defined in detail and additional information is required. In particular, information is required to define:

- farms' capital restrictions (liquidity): equity capital, asset value, land value
- production activities: revenues, variable costs, percentage of variable costs bound during a production period, (coupled) subsidies, technical coefficients on factor use (feeding requirements, labour demand, nitrogen production/uptake, average annual milk yield per cow) and crop rotation constraints
- investment options: investment costs, typical share of equity bound in investments, size/capacity of the investment, useful life, average work requirement per unit, estimates on maintenance costs
- financing activities: interest rates for long-term and short-term borrowed capital and savings
- other activities (if specific to regional structure): quota lease, manure import/export, regional ceilings, e.g. on livestock density
- labour activities: wages of farm-labour, wages of off-farm labour.

6.4 Lessons/Experiences

To parameterize AgriPoliS, available statistical and survey data is used to identify the agent types and to create the artificial population. In the end, the crucial part in the parameterization is the data availability and quality. AgriPoliS has already been applied to 22 different regions in 11 countries in the European Union. Thereby, numerous problems had to be faced regarding the data.

1. Regional data are not always available from the same source or for the same year. For example the total area of agricultural land coming from the national census might differ to the total area of agricultural land calculated by the farm census, because in the farm census only farms above a specific size are considered. For the up-scaling such discrepancies have to be identified and harmonized.
2. There may be discrepancies as well between farm level data and regional statistics. For instance, whereas the farm sample from the FADN includes in some regions no farm smaller than 20 ha, the regional statistics might include them. Ignoring those farms means that all other regional characteristics have to be recalculated, e.g. the total number of farms has to be reduced as well as the total area, the total number of animals in the region, the number of farms per size class or farm type, etc.

3. A pre-selection of regionally-relevant farms from the FADN has to be done beforehand. For instance, the FADN differentiates farms according to their technical orientation, e.g. a farm breeding suckler cows is a grazing livestock farm. However, a farm with sheep is classified as grazing livestock farm as well. The up-scaling procedure might therefore select a sheep farm instead of a suckler cow farm even though sheep are not planned to be considered in the model. When this happens the up-scaling procedure has to be done again after removing the sheep farm.

The quality of the up-scaling procedure can be evaluated by the value of the regional goal criteria, namely how small is the sum of the squared deviation of the individual regional characteristics. The closer this value is to zero, the better is the representation of the region. However, one can also check the relative deviation of individual characteristics of the virtual region from the empirical values.

6.5 Alternatives

If no farm level data (FADN or IACS data) is available it is possible to follow the sequence outlined for case 13. Thereby, agent types are created by expert knowledge, i.e., instead of taking farms from FADN or IACS, experts can define a set of farms typical for the study region which are then up-scaled applying the same method as described before. For the definition of typical farms see, e.g. Hemme et al. (1997) or Berg et al. (1997).

At the moment farmers' behaviour in AgriPoliS is economic but myopic or boundedly rational and non-strategic. However, in reality farmers may behave strategically (in a game theoretic sense) and make decisions according a specific strategy or propositions like risk aversion. To determine such strategic behaviour one can follow the sequence outlined for case 16 (because census data about the strategic behaviour of farms is not available). Furthermore, we expect only a small range of different types of strategic behaviour which might be identified by small-scale field work. A further option are behavioural experiments. Because a model is already available, one could replace a computer agent by a real person to observe the behaviour and performance of this person. The advantage of such experiments compared to field work is that they can be replicated with different framework conditions (policy changes, price volatility) or initial situations (e.g. different farms). In a next step, the observed behaviour has to be transferred into heuristics that can be implemented in AgriPoliS. Possibly, different behaviour can be observed. To assign the different behaviour to the population proxy data for the distribution of different strategic behaviour can be used.[10]

[10] The idea of behavioural experiments becomes actually (2010–2013) realized in the project "Structural Change in Agriculture" funded by the German Research Foundation (DFG).

References

Arthur WB (1989) Competing technologies, increasing returns and lock-in by historical events. Econ J 99:116–131

Balmann A (1997) Farm-based modelling of regional structural change: a cellular automata approach. Europ Rev Agr Econ 24(1):85–108

Balmann A, Odening M, Weikard H-P, Brandes W (1996) Path-dependence without increasing returns to scale and network externalities. J Econ Behav Organ 29(1):159–172

Balmann A, Lotze H, Noleppa S (1998) Agrarsektormodellierung auf der Basis 'typischer Betriebe'. Teil 1: Eine Modellkonzeption für die neuen Bundesländer. Agrarwirtschaft 47(5):222–230

Berg E, Nellinger L, Brühl C (1997) Betriebliche Auswirkungen unterschiedlicher agrarpolitsicher Szenarien n ausgewählten MOE- und EU-Ländern. Achtunddreißigste Jahrestagung für Wirtschafts- und Sozialwissenschaften des Landbaues vom 6. bis 8. Oktober 1997 Freising-Weihenstephan

Chavas J-P, Chambers RG, Pope RD (2010) Production economics and farm management: a century of contributions. Amer J Agr Econ 92(2):356–375

Grimm V, Berger U, Bastiansen F, Eliassen S, Ginot V, Giske J, Goss-Custard J, Grand T, Heinz S, Huse G, Huth A, Jepsen JU, Jørgensen C, Mooij WM, Müller B, Pe'er G, Piou C, Railsback SF, Robbins AM, Robbins MM, Rossmanith E, Rüger N, Strand E, Souissi S, Stillman RA, Vabø R, Visser U, DeAngelis DL (2006) A standard protocol for describing individual-based and agent-based models. Ecol Model 198:115–126

Grimm V, Berger U, DeAngelis DL, Polhill G, Giske J, Railsback SF (2010) The ODD protocol: a review and first update. Ecol Model 221:2760–2768

Happe K (2004) Agricultural policies and farm structures—agent-based modelling and application to EU-policy reform. Studies on the agricultural and food sector in Central and Eastern Europe, Vol. 30, IAMO

Happe K, Kellermann K, Balmann A, (2006) Agent-based analysis of agricultural policies: an illustration of the agricultural policy simulator AgriPoliS, its adaptation and behaviour. Ecol Soc 11(1), Art. 49. http://www.ecologyandsociety.org/vol11/iss1/art49/

Hazell PBR, Norton RD (1986) Mathematical programming for economic analysis in agriculture. Macmillan, New York

Hemme T, Isermeyer F, Deblitz C (1997) TIPI-CAL Version 1.0: Ein Modell zur Politik- und Technikfolgenabschätzung für typische Betriebe im internationalen Vergleich. Arbeitsbericht 2/97 des Instituts für Betriebswirtschaft, FAL, Braunschweig

Henrichsmeyer W, Cypris C, Löhe W, Meudt M, Sander R, Sothen F van (1996) Entwicklung des gesamtdeutschen Sektormodells RAUMIS96. Endbericht zum Kooperationsprojekt. Forschungsbericht für das BMELF (94 HS 021). Bonn/Braunschweig. http://www.faa-bonn.de/RAUMIS96.pdf

Hofreither MF, Kniepert M, Morawetz U, Schmid E, Weiss F (2005) Ein regionalisiertes Produktions- und Einkommenssimulationsmodell für den österreichischen Agrarsektor—Vorläufiger Endbericht. Forschungsprojekt Nr. 1319 im Auftrag des BMLFUW. Wien. http://www.dafne.at/prod/dafne_plus_common/attachment_download/45ec83878a2fb2af64851b4756bded86/1319.pdf

Jacobs A (1998) Paralleler Einsatz von Regionen- und Betriebsgruppenmodellen in der Agrarsektoranalyse. Schriftenreihe des Bundesministeriums für Ernährung, Landwirtschaft und Forsten, Angewandte Wissenschaft. Köllen. Bonn

Jayet PA, Debove E, Kleinhanss W, Küpker B, Júdez L, Xepapadeas A (2007) Land market and Genedec. Presentation at DG agriculture in Brussels, March 5th, 2007. http://www.grignon.inra.fr/economie-publique/genedec/publi/DGAGRI_PAJ.pdf

Kellermann K, Happe K, Sahrbacher C, Balmann A, Brady M, Schnicke H, Osuch A (2008) AgriPoliS 2.1—model documentation. Technical report. IAMO. Halle, Germany. http://www.agripolis.de/documentation/agripolis_v2-1.pdf

Kleingarn A (2002) Anpassungskosten von Agrarpolitikänderungen—Erstellung und Kalibrierung der Datengrundlage für ein räumlich dynamisches Multiagentenmodell der Region Hohenlohe. Diplomarbeit, Humboldt Universität zu Berlin

Sahrbacher C (2011) Regional structural change in European agriculture—effects of decoupling and EU accession. Studies on the agricultural and food Sector in Central and Eastern Europe, Vol. 60, IAMO

Sahrbacher C, Happe K (2008) A methodology to adapt AgriPoliS to a region. Technical report. IAMO. Halle, Germany. http://www.agripolis.de/documentation/adaptation_v1.pdf

Sahrbacher C, Sahrbacher A, Kellermann K, Happe K, Balmann A, Brady M, Schnicke H, Ostermeyer A, Schönau F (2012) ODD-protocol of AgriPoliS. Technical report. IAMO. Halle, Germany. http://www.agripolis.de/documentation/ODD_AgriPoliS.pdf

Simon HA (1955) A behavioral model of rational choice. Q J Econ 69:99–118

Simon HA (1956) Rational choice and the structure of the environment. Psychol Rev 63:129–138

Simon HA (1996) The sciences of the artificial, 3rd edn. The MIT Press, Cambridge

Chapter 7
The Parameterisation of Households in the SimPaSI Model for East Kalimantan, Indonesia

A. Smajgl and E. Bohensky

7.1 Modelling Description

This Chapter is based on participatory research we conducted in collaboration with the Government of Indonesia (Smajgl 2010; Smajgl and Bohensky 2012). The participatory modelling process aimed for facilitating a learning experience that involved three tiers of governance. The participatory process included the co-design of the research proposal by stakeholders who were actively involved in its implementation process by carrying out many research tasks. This was enabled by substantial capacity building activities, for instance in agent-based modelling (Smajgl 2010). Following the categories of participation suggested by Barreteau et al. (2010) our process falls into the sixth category of participatory research processes, co-building and control over model use.

The specific context was defined by a set of poverty-alleviation policies, including central Government plans for fuel subsidies and poverty cash payments. In late 2008, a fuel price reduction was discussed by the Government of Indonesia as a means to reduce poverty. Price levels were about IDR 6,000 per litre of petrol and a discussion began on how poverty could be reduced and what an effective reduction would be. The study area that was selected by the Indonesian Government includes the six southern districts of East Kalimantan, an area of approximately 220,400 km^2. Some 2 million people live in this region and there is high diversity among households which represent a wide range of urban, peri-urban, and rural livelihoods based on the primary, secondary and tertiary economic sectors.

The SimPaSI model simulates processes spatially explicitly considering locations for all entities, including households. Based on best available GIS data livelihoods options are specified. For instance, the presence of forests or water bodies in

A. Smajgl (✉) · E. Bohensky
CSIRO Ecosystems Sciences, Townsville, Australia
e-mail: alex.smajgl@csiro.au

E. Bohensky
e-mail: erin.bohensky@csiro.au

A. Smajgl, O. Barreteau (eds.), *Empirical Agent-Based Modelling – Challenges and Solutions*, DOI 10.1007/978-1-4614-6134-0_7,

the immediate surroundings of a household allows household members to engage in logging, the collection of non-timber forest products, or fishing. Additionally, changes in rainfall vary spatially depending on historic data and user assumptions for respective simulations. Thus, spatial characteristics influence heterogeneity of household agents, which is a fundamental mechanism in many agent-based models (Loibl and Toetzer 2003; Brown et al. 2004; Parker and Meretsky 2004; Parker and Filatova 2008; Almeida et al. 2010; Anselme et al. 2010; Filatova et al. 2010; Haase et al. 2010; Lagabrielle et al. 2010; Moglia et al. 2010; Perez and Dragicevic 2010; Simon and Etienne 2010). As a result, policy implications are likely to vary spatially as Sect. 4 highlights. The same policy intervention is likely to have diverse impacts due to behavioural diversity is spatially distributed and the diverse environmental conditions. Stakeholders emphasised the need for a spatially disaggregated view to allow for analysing how poverty in the highly diverse villages of East Kalimantan is likely to respond to the relevant macro policy interventions. A critical factor for stakeholders was to avoid poverty hot spots, requiring spatially explicit simulations.

The following model description of the SimPaSI model for East Kalimantan is based on the ODD protocol (Grimm et al. 2006, 2010):

Purpose The purpose of this agent-based model is to simulate the impact of fuel price changes and cash payments to poor households on poverty and use levels of natural resources. This informed central Government agencies in relevant decision making processes. Additionally, local government strategies had to be simulated, which include

- changes in logging activities,
- additional mining concessions,
- extension of oil palm plantations.

State Variables and Scales Households are a main entity, which are simulated with attributes such as number of members, income, and address. Individuals belong to households and are represented with their livelihoods, income, and education. Spatial entities are defined by overlaying digital elevation information and sub-catchment delineations with soil data, land cover data and administrative boundaries. Spatial entities have land use, surface water and groundwater as their main attributes. Environmental entities include trees, fruit, rubber, rattan, fish, honey, dolphins, deer and hornbills, as these variables were identified by Indonesian research partners as important to households in East Kalimantan.

Process Overview and Scheduling The model works in daily time steps. First, rainfall occurs according to historic rainfall data, which was provided by the Province Government of East Kalimantan. Surface and groundwater flow are calculated based on best available data for elevation, soil, and land cover. Growth of flora variables is calculated based on surface water and groundwater levels. In weekly time steps household processes take place (see next Section for the parameterisation of households and their behaviour). Household members' age is updated and their livelihoods are activated and return income for each household member. Natural resource dependent livelihoods return income

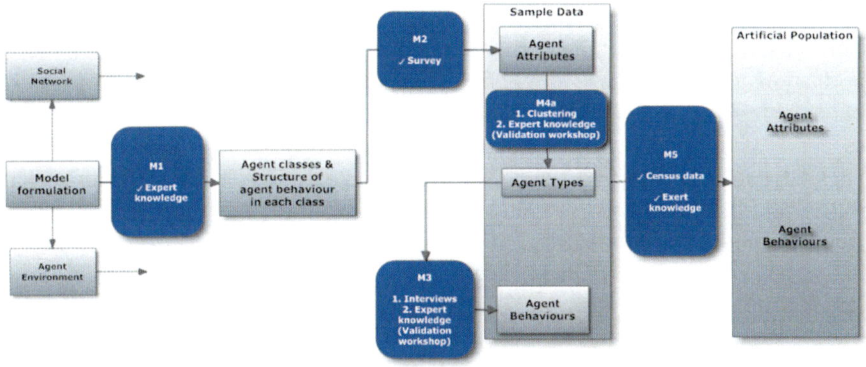

Figure 7.1 Framework for parameterisation of ABM. (Adapted from Smajgl et al. 2011a)

that depends on the state of the environmental entity, i.e. the availability of fish. Then, household income is calculated and compared with the official poverty line. If households remain under the poverty line and household members have sufficient time for a new livelihood, livelihood options are explored. Depending on the state of natural resources and on the availability of paid labour (mining, logging, plantations) new livelihoods are assigned.

Design Concept Emergent phenomena include the level of poverty and the state of natural resource stocks. Households adapt to changes they perceive by changing livelihood strategies or by migration. Households' objective is to satisfy their income goal. Households are reactive and do not formulate predictions apart from expectations on income streams when changing livelihood strategies.

Initialisation Spatial entities are initialised with official data for land use, elevation, soil type, and administrative boundaries. Rainfall is initialised with historic rainfall data. Population size, livelihoods and income are initialised with disaggregated household-level census data. Parameterisation of behavioural attributes is described below.

Input and Submodels Key submodels include households, hydrology, fish, crops and trees. Parameter values (i.e. rainfall, crop prices) are defined as range values introducing high levels of stochasticity. Parameter values build on primary data and expert knowledge (Smajgl et al. 2009b).

7.2 Characterisation and Parameterisation Methods

Our work in East Kalimantan is an example for case 4 identified in Chap. 1 of this Volume. The model formulation (M1) was mainly informed by expert knowledge, allowing for the identification of principal agent classes and principal agent behaviours as outlined by Fig. 7.1.

Households' responses to energy related policy changes were parameterised in six steps. First, a survey elicited information on household characteristics (livelihoods, motivation, assets and other socio-demographic data), which were assumed to correlate with behaviour (Trébuil 1997). A total of 3,000 households were surveyed. In a second step this data was processed in a cluster analysis. The resulting household types were in a third step presented to experts and stakeholders for validation. In a fourth step the behavioural response data was elicited for each household type by conducting in-depth interviews with a total of 512 households. This was followed by another validation workshop. This process concluded the elicitation of sample data for household attributes and household behaviour and intentions. The underpinning assumption for this sequence is that every household that does the same (= livelihood) for the same reason (= motivation) while the same constraints (= assets) will respond similarly (= intention) to the same change (= policy scenario).

The scaling-up from sample data to the target system population was based on household level census data. As types were explicitly developed disproportional up-scaling was possible. Disproportional up-scaling means that the ratio between agent types in the (interview) sample does not match the ratio between agent types in the target system population. Figure 7.1 depicts the methodological sequence. The following explains each step in detail.

7.3 Parameterisation Details and Resulting Agent Types

The household survey incorporated 27 questions on household characteristics, in particular households' livelihoods and the non-market values they believed they derive from natural and social resources (Table 7.1). Twelve of these questions asked households for type and level of benefit generated from 17 natural and social resources. The objective was to develop a typology based on key characteristics that allow distinguishing households in East Kalimantan, their livelihood strategies, and their values, in order to 'statistically predict' their intended response to particular policy or economic changes at the broader (national or regional) level. We assumed that households that implement similar livelihood strategies (current behaviour) and follow similar motivations (expressed by how they value natural and social resources) will show similar responses to these changes (Trébuil et al. 1997; Byron and Arnold 1999; Bohensky et al. 2007; Smajgl et al. 2007).

Data were collected on household location, composition, assets, wage income, and benefits derived from natural and social resources (Table 7.2). The survey was carried out by a local research team at approximately 3,000 households spread equally across four *kabupaten* or districts, and two *kota*, or municipalities.

In the second step, a cluster analysis was performed to determine household typologies (see Herr 2010 for technical details). The final typologies depended on an overall set as well as site-specific sets of clusters. Considering that most of the variables were categorical, ranking variables according to their explanatory power for identifying clusters was derived from decision trees (with a limited set of variables)

Table 7.1 Categories of survey questions

Household identification & location	Household composition	Assets	Wage income	Benefits from natural & "social" resources
Name of household head	Identity of respondent (e.g. role in household)	Number of assets owned (e.g. house, car, motorbike, fishing boat)	Who earns	Type of use or value of natural resources
Address	size	Assets owned that are worth more than annual salary	Type of work	Type of use or value of social resources (education, roads, recreation areas, social networks)
District	Demographics		Location of work	Frequency of use
Village	Education		Time spent working	Distance travelled to use
Type of house	Origin		Daily wages	Mode of transport to use
	Ethnic group(s)			Importance for income, nutrition, health, cultural values, recreation, security

through which a typology could be assigned to each household in the area. The cluster analysis identified 19 household types.

In a third step, the household types and their statistical characteristics were presented to our stakeholders and local experts. During a three-day workshop stakeholders confirmed the resulting household types and added necessary context to each type. Table 7.2 lists the workshop results for the four household types that emerged for the district of Kutai Kartanegara as an example.

After the household types were validated semi-structured interviews were carried out, again by the local research team, in order to identify intended responses to eight policy scenarios, including fuel price increases (Smajgl et al. 2009a). A total of 540 households were interviewed, ninety households at each of the six subdistricts. The first interview section listed a series of questions corresponding to the decision tree variables to identify the household type (Herr 2010). Next, eight hypothetical scenarios were described related to energy policy change or employment opportunities. For each scenario households were asked how this would affect a household's use of natural resources, the hours of paid work it undertakes per week, migration with and without the rest of the household, investment in assets (i.e. a motorbike, house or boat), and application for work should a new coal mining, logging, or oil palm company begin operating in the area. Questions related to migration or new work asked about likely location to spatially reference the intended behaviour. Households were also asked if they would do anything differently that was not already specified. In a final set of open-ended questions households

Table 7.2 Example of four household types for the district of Kutai Kartanegara

Timber users	Migrants	Fishermen	Forest dwellers
Second largest group (60/440)	Largest group in Kartanegara (283/440)	Smallest group (44/440)	Third-largest group (53/440)
Primary and secondary sectors dominate employment	Below average income	Low income group	Mostly employed in primary industries
Vast majority has valuable assets	Majority has no valuable assets	Employed in fishing	22 % in poorest group and average around Rp1.5m
25 % value timber for income	Do not value social networks for income	100 % value fish for income	35 % value wild pig and kijan for income (hunting)
Nearly all (95 %) social networks and education	Do not own boat engine	40 % value education for income	Most (80 %) value social networks for income
Majority (55 %) value roads for income	Do not value fish or timber for income	Most do not value social networks (78 %) or roads for income (70 %)	Most (92 %) value education and roads for income
Represents households that live in suburbs/villages close to Tenggarong with good access to markets, old transmigration area, most developed area	Represents migrants/newcomers, who mostly live off illegal logging in upper parts of area, in squatting areas or new development	Represent fishermen that live in more accessible areas (Kecamatan Kota Bangun & Danau Jempan, Muara Wis), with good transportation and market access (buyers come)	Nobody has valuable assets
Represents households that sell processed/sawn timber			Represents very traditional remote villages along Mahakam Riverbank (Belayan River)
			Some households earn significant amounts from illegal logging

Variables shown are those with the highest discriminatory power

Table 7.3 Categorisation of interview results for policy scenario #1 (fuel subsidy reduction) for workshop validation

Question	Response	Household types
Change natural resource use	None (<20 %)	PPU "Immigrants" & "manufacturing", Kutai Kartanegara "timber users", Paser "Hinterland dwellers", Kutai Barat "transmigration area", Balikpapan "urban centre" & "urban poor", Samarinda "environmentalists"
	Increase (≥20 %)	Kutai Kartanegara "migrants"
Change hours of paid labour	No change	Kutai Barat "traditional" & "transmigration area", Balikpapan "urban centre" & "urban poor"
	Increase (≥20 %)	PPU "immigrants" & "manufacturing", Kutai Kartanegara "timber users" & "migrants", Paser "Hinterland dwellers" & "farmers", Samarinda "environmentalists"
Migration	No change (<20 %)	PPU "manufacturing", paser "Hinterland dwellers" & "farmers", KuBar "traditional", Balikpapan "urban centre" & "urban poor", Samarinda "environmentalists"
	Migration out (≥20 %)	PPU "immigrants", Kutai Kartanegara "timber users" & "Migrants", Kutai Barat "transmigration area"

Codes listed in the "Household types" column represent the types found in the six study sites Feedback (right-hand column) was captured and entered interactively with workshop participants

were given the opportunity to elaborate on their earlier responses and to allow for cross-checking of consistency.

The sixth step included another validation workshop, in which interview results were presented to regional experts and stakeholders in order to validate responses, and to define and clarify agent rules. Table 7.3 summarises the key results of the interviews and the second validation workshop.

In the final step (M5) household level census data was used for assigning behavioural types to each household in the region and, thereby, assigning a spatial distribution of household types. Technically, main characteristics of household types and data points in the census data provided sufficient overlap to allow for mapping types into the census data. The census data provided spatial references down to the village level and ensured a realistic population number for each village. Therefore, intended behaviour could be spatially distributed without keeping the proportions of household types captured in the sample. This kind of disproportional up-scaling is required if the initial survey is likely to misrepresent realistic proportions of household types (Smajgl et al. 2011a). Proportional up-scaling (i.e. cloning) would lead to overrepresentation of minorities, see Smajgl et al. (2011a).

7.4 Lessons Learnt

The overarching objective of the modelling process was to facilitate a learning process among decision makers from three levels of governance. This task required capturing and mapping household behaviour in a diverse region. Capturing intentional data according to its spatial occurrence allowed us considering likely behavioural responses when analysing large scale policy changes. Critical for simulation outcomes was the spatial location of behavioural responses; evenly spread changes in the use of natural resources would have impacted the social-ecological system differently from a situation with spatially concentrated responses. A benefit of the parameterisation technique was the time- and cost-savings, in that the cluster analysis allowed us to reduce number of in-depth interviews to 540. However, there were also a few limitations. One is that the cluster analysis did not automatically provide unambiguously interpretable variables, requiring expert workshops for contextualisation and validation. While our process resulted in stakeholders and experts immediately recognising household types it seems not unlikely that the clustering would require a few iterations before finding types that can be validated. This adds time to the parameterisation process, and while this engagement and involvement are critical aspects of the participatory nature of this research, it also increases the possibility of shifting results closer to stakeholder expectations. Such biases diminish the strength of evidence driven model implementation and, thereby, the potential for learning. Additionally, because this was a previously untested approach, a rigorous pilot study to test the entire process would have been beneficial to identify ambiguity of particular questions, language and interpretation issues, and data entry problems. Most of these issues were easily resolved in consultation with stakeholders.

A similar participatory process was conducted in Central Java, also employing agent-based modelling to facilitate a learning process (Smajgl et al. 2009a). However, the parameterisation process in Central Java was conducted without developing explicit agent types. Instead, proportional upscaling was implemented (case 1 in Fig. 1.2 of Chap. 1). Camparing these two approaches requires explicit evaluation criteria. Considering the goal to facilitate a learning process stakeholder responses seem paramount to this evaluation. Based on stakeholder responses on the validity of modelling results, it seems that the rigorous approach described above does not add substantially to model robustness if compared to proportional up-scaling methods (i.e. if a sample with intentional data is assumed to be representative and therefore each respondent is multiplied—or cloned—until the whole population is initialised). The most critical system property for deciding what method to apply is the relevance of behavioural minorities. If the study entails dynamics driven by minorities proportional up-scaling based on sample data is likely to either overestimate the minority (in case one or a few representatives are captured in the sample) or underestimate the presence of the critical behaviour (in case the sample does not include such behaviour). Thus, in cases with critical minorities disproportional up-scaling seems superior to proportional up-scaling. However, disproportional up-scaling requires an additional database (i.e. disaggregated census data or GIS data) and robust assumptions for mapping sample-based behaviour into the disaggregated database, which introduces additional uncertainties that can hardly be quantified. For instance, if household types are up-scaled based on a few socio-demographic characteristics that can be found in census data the process is technically feasible. But if these data points allow for a robust and meaningful mapping of types into the whole population is critical. Technical feasibility does not automatically ensure robustness of the parameterisation process. Thus, comparing proportional up-scaling (case 1) with disproportional up-scaling (i.e. case 4) requires an explicit evaluation of the improved accuracy of proportions of household types in the target population. Types are introduced to avoid unrealistic proportions. If the up-scaling cannot ensure a superior proportionality the additional steps (i.e. clustering) seems unjustified.

In summary, the method showcased in this paper reduces some uncertainties (i.e. minority problem) but introduces other uncertainties (i.e. mapping behaviour onto census data). This comparison has to remain qualitative as approaches for evaluating parameterisation techniques are not well developed or tested. Model evaluation requires robust metrics and the validation debate in the agent-based community does not seem sufficiently matured (see Smajgl et al. (2011b) for discussion and examples). Therefore, the question regarding how far specific parameterisation techniques can contribute to the 'realism' or 'robustness' of an empirical agent-based model cannot yet be addressed and parameterisation options cannot be easily compared.

Nevertheless, this Chapter tested and discussed a sequence of methods for the type of ABM that aims to simulate a large population with complex behavioural characteristics. Ultimately the process has broad applicability to agent-based modelling that strives to incorporate realistic human behaviour through empirical data.

References

Almeida SJd, Martins Ferreira RP et al. (2010) Multi-agent modeling and simulation of an Aedes aegypti mosquito population. Environ Model Softw 25(12):1490–1507

Anselme B, Bousquet F et al (2010) Modelling of spatial dynamics and biodiversity conservation on Lure mountain (France). Environ Model Softw 25(11):1385–1398

Barreteau O, Bots PWG et al (2010) A framework for clarifying "Participation" in participatory research to prevent its rejection for the wrong reasons. Ecol Soc 15(2):1

Bohensky E, Smajgl A et al (2007) Calibrating behavioural variables in agent-based models: insights from a case study in East Kalimantan, Indonesia. international congress on modelling and simulation (MODSIM07), Virginia Beach, VA modelling and simulation society of Australia and New Zealand

Brown DG, Page SE et al (2004). Agent-based and analytical modeling to evaluate the effectiveness of greenbelts. Environ Model Softw 19(12):1097–1109

Byron N, Arnold M (1999) What futures for the people of the tropical forests? World Devel 27(5):17

Filatova T, Voinov A et al (2010) Land market mechanisms for prevention of space for coastal ecosystems: an agent-based analysis. Environ Model Softw 26(2):179–190

Grimm V, Berger U et al (2006) A standard protocol for describing individual-based and agent-based models. Ecol Model 198(1–2):115–126

Grimm V, Berger U et al (2010) The ODD protocol: a review and first update. Ecol Model 221(23):2760–2768

Haase D, Lautenbach S et al (2010). Modeling and simulating residential mobility in a shrinking city using an agent-based approach. Environ Model Softw 25(10):1225–1240

Herr A (2010) Statistics for categorical surveys-a new strategy for multivariate classification and determining variable importance. Sustainability 2:533–550

Lagabrielle E, Botta A et al (2010) Modelling with stakeholders to integrate biodiversity into land-use planning—lessons learned in Réunion Island (Western Indian Ocean). Environ Model Softw 25(11):1413–1427

Loibl W, Toetzer T (2003) Modeling growth and densification processes in suburban regions–simulation of landscape transition with spatial agents. Environ Model Softw 18(6):553–563

Moglia M, Perez P et al (2010) Modelling an urban water system on the edge of chaos. Environ Model Softw 25(12):1528–1538

Parker DC, Filatova T (2008) A conceptual design for a bilateral agent-based land market with heterogeneous economic agents. Comput Environ Urban Syst 32(6):454–463

Parker DC, Meretsky V (2004) Measuring pattern outcomes in an agent-based model of edge-effect externalities using spatial metrics. Agri Eco Environ 101(2–3):233–250

Perez L, Dragicevic S (2010) Modeling mountain pine beetle infestation with an agent-based approach at two spatial scales. Environ Model Softw 25(2):223–236

Simon C, Etienne M (2010) A companion modelling approach applied to forest management planning. Environ Model Softw 25(11):1371–1384

Smajgl A (2010) Challenging beliefs through multi-level participatory modelling in Indonesia. Environ Model Softw 25(11):1470–1476

Smajgl A, Bohensky E (2012) When households stop logging—evidence for household adaptation from East Kalimantan. For Policy Econ 20(0):58–65

Smajgl A, Bohensky E et al (2007) Capturing heterogeneity in empirical agent-based models: a guideline. Fifth ESSA conference, Toulouse

Smajgl A, Carlin G et al (2009a) Assessing impacts of fuel subsidy decisions on poverty and fish catch in Central Java, Indonesia: an agent-based analysis. analysing pathways to sustainability in Indonesia: working Paper 4. Townsville, CSIRO sustainable ecosystems

Smajgl A, Carlin G et al (2009b) Simulating pathways to sustainability in Indonesia: design document for agent-based model SimPaSI East Kalimantan. Townsville, CSIRO

Smajgl A, Brown DG, Valbuena D, Huigen MGA (2011) Empirical characterisation of agent behaviours in socio-ecological systems. Environ Model Softw 26: 837–844.

Smajgl A, House A et al (2011b) Implications of ecological data constraints for integrated policy and livelihoods modelling: an example from East Kalimantan, Indonesia. Ecol Model 222:888–896.

Trébuil G, Kam SP et al (1997) Systems diagnoses at field, farm and watershed levels in diversifying upland agroecosystems: towards comprehensive solutions to farmers' problems. Applications of systems approaches at the farm and regional levels. Kluwer Academic, The Hague, pp 99–114

Chapter 8
Parameterisation of Individual Working Dynamics

S. Huet, M. Lenormand, G. Deffuant and F. Gargiulo

How do European rural areas evolve? While for decades the countryside in many regions of Europe was synonymous with inevitable decline, nowadays, some areas experience a "rebirth, even in areas where until recently development was not considered possible" (Champetier 2000). A recent EPSON (European Observation Network for Territorial Development and Cohesion) project report (Johansson and Rauhut 2007), concludes that "since the 1970s a global process of counter-urbanization has become increasingly manifest". However, this general rebirth of the countryside hides deep heterogeneities. That can be observed in the Cantal "département" in France where the population remains stable after having been depopulated with some subgroups of its municipalities have an increasing population while others have a decreasing one. Our modelling effort aims at better understanding these heterogeneities.

Micro modelling (Gilbert and Troitzch 2005) is a very relevant paradigm to study the evolution of areas composed from various objects appearing as very heterogeneous. It includes three different approaches: cellular automata change (Ballas 2007, p. 17, 2005, p. 3, 2006 p. 4; Brown et al. 2006, p. 18; Coulombel 2010, p. 66; Moeckel 2003, p. 54; Rindfuss 2004, p. 20; Verburg 2004, p. 11, 2006, p. 8, 2002, p. 12), microsimulation (Orcutt 1957, p. 287; INSEE 1999, p. 2; Holme 2004, p. 53; Turci 2010, p. 70; Morand 2010, p. 71) and agent-based models (Deffuant 2008, p. 283, 2005, p. 9, 2002, p. 7, 2001, p. 36; Bousquet 2004, p. 202; Brown and Robinson 2006, p. 63; Fontaine and Rounsevell 2009, p. 65;

S. Huet (✉) · M. Lenormand · G. Deffuant · F. Gargiulo
Laboratoire d'Ingénierie des Systèmes Complexes, Irstea, Aubiere, France
e-mail: sylvie.huet@irstea.fr

M. Lenormand
e-mail: maxime.lenormand@irstea.fr

G. Deffuant
e-mail: guillaume.deffuant@irstea.fr

F. Gargiulo
e-mail: Floriana.gargiulo@gmail.com

A. Smajgl, O. Barreteau (eds.), *Empirical Agent-Based Modelling – Challenges and Solutions*, DOI 10.1007/978-1-4614-6134-0_8,

Parker Dawn 2003, p. 9) which have been already used to study problem close to ours. However, recent reviews recommend a hybrid approach (Birkin and Clark 2011, p. 81; Birkin and Wu 2012, p. 90), particularly coupling microsimulation and agent-based modelling. Thus, trying to develop an approach which is as close to the data as we can, we decide to use microsimulation and agent approaches allowing us to address some complex individual dynamics, largely unknown and for which no data are available, such as the residential location decision (Coulombel 2010, p. 66).

The problem of such modelling approach is the link to data. If it is obvious in the basic microsimulation, that is not so easily manageable in dynamic microsimulation with a "real" evolution time after time of the individual. Indeed the dynamic microsimulation remains rare (Birkin 2012, p. 90): the most common way to introduce change of the demographic structure is to apply static ageing techniques consisting in reweighting the age class according to external information. That is to avoid considering functions of evolution of the behaviour of the individual and their parameterisation. Regarding the multiagent modelling, (Berger and Schreinemachers 2006) argue it "holds the promise of providing an enhanced collaborative framework in which planners, modellers, and stakeholders may learn and interact. The fulfilment of this promise, however, depends on the empirical parameterization of multiagent models. Although multiagent models have been widely applied in experimental and hypothetical settings, only few studies have strong linkages to empirical data and the literature on methods of empirical parameterization is still limited." An example can be read in (Fernandez 2005, p. 64) which initialise individual preference from analyses of the data coming from an ad hoc survey but don't consider a possible change in the preference of an individual.

In our model,[1] we tried to have a strong linkage to data both in the definition of the initial population and the one of the individual behaviour. This model implements virtual individuals, members of households located in municipalities and their state transitions corresponding to demographic and changing activity events: birth, finding a partner, moving, changing job, quitting their partner, retiring, dying …. The virtual municipalities offer jobs and dwellings which constrain the possible state transitions. Because we are interested in understanding better the dynamics leading to the development or, on the contrary, to the decline and possible disappearance of municipalities and settlements, two sets of cruxes can be identified in the model: The individual dynamics which determine the needs for residence and jobs; the dwelling and the job offers exogenous and endogenous dynamics at the local (i.e. municipality) level.

The present paper focuses on how to make such a model close enough to the data to guarantee a good understanding of the dynamics of population/depopulation based on "real" situations, and a real utility for policy makers. As the developed model is very large, taking into account many dynamics, we are going to

[1] This work has been funded under the PRIMA (Prototypical policy impacts on multifunctional activities in rural municipalities) collaborative project, EU 7th Framework Programme (ENV 2007-1), contract no. 212345.

focus on the design and the parameterisation of the individual dynamics regarding the labour market.

After a summary of the whole model, presented in details in (Huet et al. 2011, p. 189), we present how we have conceived and parameterised the submodel of the individual activity dynamics. The final section tries to explain what we have learnt from such an exercise. In particular, we want to stress out the necessity not to only consider the objectives of the model during the design phases, but also since the very beginning censing the existing data sources and studying the implicit model beside the databases.

8.1 Model Description

We have adopted a micro-modelling approach. The presentation of the model globally follows the requirements of the ODD (Overview, Design concepts, and Details) framework (Grimm et al. 2006). Indeed, this recently updated protocol (Grimm et al. 2010) has proved its utility to describe properly complex individual-based models, for example in (Polhill et al. 2008, p. 282).

The purpose of the model is to study how the population of rural municipalities evolves. We assume that this evolution depends, on the one hand, on the spatial interactions between municipalities through commuting flows and service, and on the other hand, on the number of jobs in various activity sectors (supposed exogenously defined by scenarios) and on the jobs in proximity services (supposed dependent on the size of the local population). Indeed, in the literature, the most cited explanation for the evolution of the rural municipalities is what is called the residential economy (Davezies 2009, p. 73; Blanc and Schmitt 2007, p. 32). It argues that rural areas dynamics is linked to the money transfers between production areas and residence locations. These money transfers are for instance performed by commuters, or by retirees who move from the urban to the rural areas. Indeed migrations from urban to rural areas are also considered as a very important strand for rural areas evolution (Perrier-Cornet 2001, p. 24). The residential economics studies particularly how an increasing local population (and money transfers) increases the employment in local services. The geographic situation plays also a role in the municipality evolution (Dubuc 2004, p. 26). To summarise, existing literature stresses the importance of the different types of mobility between municipalities, commuting, residential mobility (short range distance), migration (long range distance) (Coulombel 2010, p. 66) and the local employment offer generated by the presence of the local population.

These two aspects have to be properly taken into account in our model, since our objective is to study through simulations the dynamics of rural areas. Obviously, it appears also essential to model the demographic evolution of the municipality considering the strands explaining the local natural balance.

8.1.1 Main Entities, State Variables and Scales

The model represents a network of municipalities and their population. The distances between municipalities are used to determine the flows of commuting individuals (for job or services). Each municipality comprises a list of households, each one defined as a list of individuals. The municipalities also include the offers of jobs, of residences and their spatial coordinates. Here is the exhaustive list of the main model entities with their main attributes and dynamics.

8.1.1.1 Municipalityset

The set of municipalities can be of various sizes. It can represent a region of type NUTS 2 or NUTS 3,[2] or more LAU or intermediate sets of municipalities such as "communauté de communes" in France. In the present paper, the set corresponds to the Cantal "département" in France composed of 260 municipalities.

Parameter a threshold distance called "proximity" between two municipalities; beyond this distance the municipalities are considered too far from each other, to allow commuting between them without considering to move for instance (parameterised at 25 km).

8.1.1.2 Municipality

It corresponds to LAU2.[3] The municipality is the main focus of the model. It includes:

- A set of households living in the municipality. The household corresponds to the nuclear family.[4] It includes a list of individuals who have an occupation located inside or outside the municipality).
- The set of jobs existing on the municipality and available for the population of the model (i.e. subtracting the jobs occupied by people living outside the modelling municipality set).
- The distribution of residences, or lodgings, on the municipality.

There is a particular municipality, called "Outside": it represents available jobs accessible from municipalities of the considered set, but which are not in the considered set. The job offer of Outside is infinite and the occupation is defined by a probability of individuals to commute outside the set (see Sect. 8.2.9 for details).

[2] Eurostat defines the NUTS (Nomenclature of Territorial Units for Statistics) classification as a hierarchical system for dividing up the EU territory: NUTS 1 for the major socio-economic regions; NUTS 2 for the basic regions for the application of regional policies; NUTS 3 as small regions for specific diagnoses; LAU (Local Administrative Units 1 and 2) has been added more recently to allow local level statistics.

[3] Consists of municipalities or equivalent units.

[4] A nuclear family corresponds to the parents and the children; that is a reductive definition of the family corresponding on the most common way to define the family in Europe nowadays.

Table 8.1 Attributes defining the household state

Name	Type	Values
Members	List of individuals	
Couple	Boolean	True, false
Leader	Individual	
Residence	Residence	
Residence need	Boolean	True, false
Municipality of residence	Municipality	

Parameters:

- An initial population of households composed of individuals with their attribute value and their situation on the labour market
- A residence offer: available number of residences for each type. A type corresponds to the number of rooms
- A job offer: number of jobs offered by the municipality for each type of job; the exogenously defined part of job offers is distinguished from the endogenously defined part in order to update this last part easily
- The laws ruling the proximity of municipalities: each municipality has rings of 'nearby' municipalities (practically every 3 Euclidian kilometres) with a maximum distance of 51 Euclidian km. The accessibility of each ring varies depending on the process (commuting, looking for a residence, looking for a partner) following appropriate probability distribution laws.
- Spatial coordinates

As said earlier, in the case of special municipality called "Outside", all variables, except job offer and job occupation, are empty.

8.1.1.3 The Job and the Residence

A job has two attributes, a profession and an activity sector in which this profession can be practiced. It is available in a municipality and can be occupied by an individual. The profession is an attribute of the individual and can take six various values (see Sect. 8.1.1.5 for details) at the same time it defines a job. There are four activity sectors: Agriculture, Forestry and Fishing; Industry; Building; Services and Commerce. Overall, considering the six professions for four activity sectors, we obtain 24 jobs to describe the whole diversity of jobs in the region we study (i.e. the Cantal "département", called only Cantal later in this chapter).

The residence has a type which is classically its size expressed in number of rooms. A residence is available in a municipality and can be occupied by 0, one or more households. Indeed several households can live in one residence for instance when a couple splits up and one of the partner remains in the common residence for a while. It is also the case in some European countries where it is customary for several generations to live under the same roof (Table 8.1).

Table 8.2 Attributes defining the state of an individual

	Type	Values
Activity status	Enum	Student, inactive, retired, employed, unemployed (only the two last can search a job)
Profession	Enum	Farmers; craftsmen, storekeepers, business owners; top executive managers, upper intellectual profession (senior executives); intermediary professions; employees; workers
Job	Couple of values	24 couples (profession, activity sector) (see Sect. 8.1.1.3 for details)
Place of work	Municipality	Nil or a municipality
Household status	Enum	Adult, child
Age to die	Integer	Drawn from a distribution
Age in labour market	Integer	Drawn from a distribution
Age of retirement	Integer	Drawn from a distribution

8.1.1.4 Household

For the initialisation, residences are associated randomly with households. Then, new households are created when new couples are formed or when people from outside the set of municipalities migrate into the municipality. Households are eliminated when their members die, or when the couple splits up, or when they simply migrate outside the municipality set. When a behavior of an individual has an impact on the household, a leader is assigned randomly, or designed depending on the process. This leader will be the one deciding for the household. That is for example the case when an individual finds a job very far: she becomes the leader to make the household moving and finding a residence close to her new job.

8.1.1.5 Individual

The individual is instantiated via one of the adults of a household having the "couple" status in the birth method, or directly from the initialisation of the population, or by immigration.

The age to die, the age the person will enter the labour market, and the age of retirement are attributed to the individual when it is created. These ages are assigned by a probability method. The activity status defines the situation of the individual regarding employment, especially whether or not she is looking for a job. The individual can quit a job, search for and change jobs ....

The profession is an attribute of the individual indicating at the same time her skills, level of education and the occupation she can aspire to. Professions take the value of the French socio-professional categories categorised in six modalities that define at the same time a kind of occupation, an average level of education and an approximate salary (Table 8.2)

8.1.2 *Process Overview and Scheduling*

8.1.2.1 The Main Loop

The main loop calls processes ruling demographic evolution, the migrations, the job changes, and their impact on some endogenously created services and/or jobs. First, the scenarios are applied to the municipalities. Then, endogenously available jobs and services are updated in municipalities. Finally, demographic changes are applied to the list of households. The following pseudo code sums-up the global dynamics:

```
At each time step:
  For each municipality
    municipality.update  external  forcings:  offer  of
      jobs, residence
    municipality.update endogenous job offer for ser-
      vices to residents
    municipality.compute in-migration
  For each household:
    household.members.job searching decision (this pro-
      cess can make free some jobs from people becoming
      retired or inactive)
  For each household:
    household.members.searching for a job
    household.members events (coupling, divorce, birth,
      death)
    household.residential migration
    household.members.individual ages
```

Time is discrete with time steps corresponding to years. The households are updated in a random order during a time step. We shall calibrate the model on the first 16 years and study its evolution on the next 24 years.

8.1.2.2 Dynamics of Offer for Jobs, Services and Lodging

In the municipality objects, jobs, services and dwelling offers are ruled. Changes in dwelling offers are specified in scenarios. Various sizes are considered in order to match the needs of households.

The job offer process is twofold: one part defined through scenarios which specify the increase or decrease of jobs in different sectors, and a second part concerning the proximity of service jobs, which are derived by a specific statistical model.

Indeed, numerous are the researches pointing out the importance of services for the rural areas dynamism (Aubert 2009, p. 22; Dubuc 2004, p. 26; Fernandez 2005, p. 64; Soumagne 2003, p. 30). Also the residential economics shows the im-

Table 8.3 Regression coefficient for the four classes of municipalities of the Cantal

Classes of distance in minutes to the most frequented municipality	β_0	β_1
0	−0.170901146	0.033121263
[0,5]	−0.130158882	0.025111874
[5,10],	−0.141049558	0.026983278
>10	−0.162030187	0.031165605

portance of the presence of the population in rural municipalities (Davezies 2009, p. 73). Practically, we distinguish the proximity services which rely directly on the presence of population from the services which are decided according to other factors (assets of the location, political will at different levels, etc.). We integrated the dynamics of creation and destruction of proximity services jobs in the micro-simulation model, using a statistical model derived from the data of the region. Starting from the classical minimum requirement approach proposed by (Ullman and Dacey 1960, p. 259; Lenormand et al. 2012a, p. 258) we propose a model which takes into account the distance between a municipality and its closest centre of services (i.e. most frequented municipality, called MFM). This new model has been grounded on detailed data related to jobs and poles of services (Lenormand et al. 2012a, p. 61). Therefore, we use the extracted statistical relation to adjust the number of jobs in proximity services in the municipalities of the model.

It is $E=\beta_0+\beta_1 \ln P+\varepsilon$ with E= minimum employment offer in the municipality to satisfy the need for services of one resident; P= the population of the municipality; β_0 and β_1= parameters

For each municipality, this function is computed every year in order to update the service sector job offer depending on the distance of the municipality to the closest pole of service (called MFM). The form of the function for different municipality sizes with various distances to the MFM indicates that:

- in any case, the job offer is higher in the pole of services and decreases in the surrounding;
- however further from the pole of services, the number of jobs increases again until reaching a plateau at a distance higher than 10 min;
- the larger is the municipality, the higher is the number of jobs in proximity services.

The other creations and destructions of jobs are ruled by scenarios.

Parameters Distances to the Most Frequented Municipality of every municipality of the Cantal (given by the French Municipal Inventory of 1999); class of distance to the most frequented municipality (MFM) for every municipality and regression coefficients β_0 and β_1 extracted of the analysis of the French Census of 1990, 1999 and 2006 (see (Lenormand et al. 2012a, p. 61) for more explanations; Table 8.3).

The proportion of proximity service jobs offer over professions is assumed to be the same than the one for the whole service sector job offers (which is probably a strong approximation). This allows us to distribute the proximity service jobs in the different jobs in the service sector.

8.1.2.3 Dynamics of Labour Status and Job Changes

A new individual can be generated in a household having the "couple" status with the birth method, or directly from the initialisation of the population, or from the immigration method. A newly born individual is initialised with a student status that she keeps until she enters the labour market with a first profession. Then, she becomes unemployed or employed with the possibility to look for a job. She may also become inactive for a while. When she gets older, she becomes a retiree. We here describe rapidly these dynamics to situate them in the global picture of them model. We describe them in more details, especially the choice of parameters and link to data, in Sect. 8.3.

8.1.3 Entering on the Labour Market

The individual stops being a student at the age to enter on the labour market and becomes unemployed. She searches immediately for a job and can get one during the same year. A first profession she looks for has to be defined at the same time the first age of research is determined.

Parameters Probabilistic laws to decide the age a student enters on the labor market and the first profession she is going to look for.

8.1.4 Job Searching Decision

The decision for searching a job is a two-step process. First, an individual has an activity status indicating if she is susceptible to search for a job or not. She can change her status and then her probability to seek a job. When she decides searching, she has also to decide what type of job to search for. Five different activity statuses define the individual situation regarding the labour market in the model:

- The **student**: an individual is a student in the first part of its life, until the age she enters on the labour market. We consider the probability of a student to look for a job is 0 since we are only interested in rural municipalities. Students in age working mainly look for a job in the large cities where they study.
- The **unemployed**: an individual is unemployed when she is considered active (on the labour market) and has no job. For sake of simplicity, we assume an unemployed has a probability 1 to look for a job.
- The **employed**: she is an individual who has a job. She can decide searching for another job, in the same profession or not. Her probability willing to change job classically depends at least on her age.

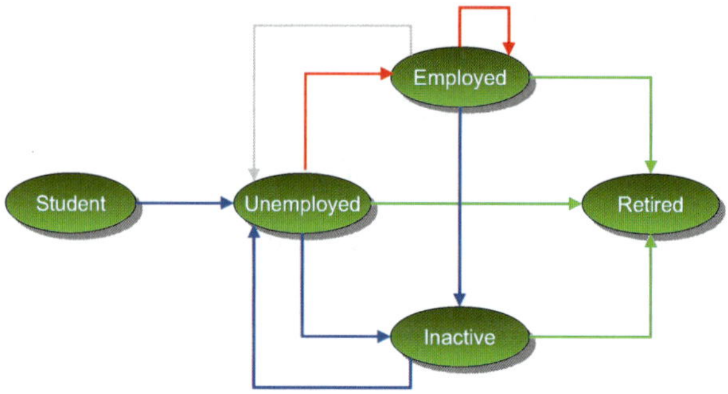

Figure 8.1 Transitions of status and their link to the data. *Red arrows*: change by finding a job; *grey arrows*: when she is fired; *green arrows*: at the age of retirement (picked out from a law extracted from data); *yellow arrows*: due to a probabilistic decision of becoming inactive extracted from the Labor Force Survey data; *purple arrows*: due to probabilistic decisions extracted from the Labor Force Survey data

- The **inactive**: she can be inactive for a long time or just stopping to work for 1 year, having a baby for example. During this period, her probability to search for a job is 0.
- The **retired**: at the age of retirement, an individual retires. Her probability to look for a job is then assumed to be 0.

We have seen the probability to search for a job (or the law ruling this probability) depends on the activity status. Figure 8.1 describes the way an individual changes activity status and thereby the probability to search.

Entering the labour market, the student becomes unemployed and searches for a job with a probability 1. An unemployed, as an employed, can find a job through processes presented in the following sections and become employed. If an unemployed always searches for a job by assumption that is not the case for an already employed individual (her probability to search has to be extracted from data). Employed and unemployed individuals can also become inactive. Then we assume that they stop searching for a job the time they remain inactive. Every activity states, except student, can be followed by the retirement state in which we assume the individual stops searching for a job. An inactive, if she doesn't retire, either can come back on the labour market adopting an unemployed status to search for a job or can remain inactive.

Most of the laws ruling the activity status changes have to be parameterised. The grey-arrows transitions are much more endogenously defined. That is the employed to unemployed transition which is due to the decreasing availability of job offer implying a sacking. It can also be, for instance a resignation of an individual leaving her municipality to follow her partner to another place of residence.

Knowing an individual searches for a job, we have to compute which profession she looks for. One can notice that an individual only looks for a profession; we neglected to take into account the activity sector in her choice. The activity sector will be defined by the found job among the set of possible job offers for the individual.

We expect the job offer to be a sufficient constraint on the activity sector to allow the model exhibiting a statistically correct distribution of occupied jobs by activity sector.

Parameters Ruling the Job Research Decision Probability becoming inactive; probability to stop being inactive; probability laws defining what profession to search for; parameters for entering the labour market and to retire.

8.1.5 Searching for a Job

The question for the individual is now to decide where to search for a job. The challenge consists in preserving the properties of the commuting distance distribution that we assume constant. Both the choice of the place of work and the choice of the place of residence impact on this distance. Thus, these processes have to be designed under this constraint. However, the place of work is not only defined by the strategy of search but also constrained by the job offer, which has to be properly defined.

If the leader of the household has already found a job far (further than the proximity attribute) from the place of residence and the household is trying to move close the leader's place of work, then the other household members, waiting for a change of residence, do not try to change job since they do not know where they will be living. Until the household finds out a new residence place, nobody is going to change jobs.

In the other cases, if the individual is searching for a job, we consider she begins by choosing where she wants to work. Practically, she picks out a distance in the probability law of the "accepted distance to work place".

Then, if the distance is higher than 0, she has to decide whether to work outside the set of municipalities. The decision to work outside is described in detail in Sect. 8.2.9. If the individual goes to work outside, she automatically has a job. She is counted as an outside commuter. The job occupation of the outside and its spatial distribution can be used to calibrate the model.

If she doesn't work outside, she goes to see the labour office. The labour office collects every job offer corresponding to the profession she is looking for at the chosen distance. Then the individual chooses one at random. This procedure allows reproducing the effect of the quantity of local offers. It gives to the municipality with a larger job offer a greater probability to be chosen.

If she chooses a job at a distance higher than the proximity distance, she becomes the leader of her household. If the distance is less than the proximity, the next household member, if she exists, will be able to search for a job. The search procedure is repeated x times if the individual has not found a job. The number of times this procedure is repeated is specified in a parameter.

Parameters Probability distribution of accepted distances to cross over to work place; probability to commute outside for an inhabitant of every municipality.

8.1.6 Become a Retiree

At a given age, the individual becomes a retiree. We assume, for sake of simplicity, that a retiree does not search for a job.

Parameter Probability to decide the individual's retirement age.

8.1.6.1 Demographic Dynamics

A new household can be created when an individual becomes an adult or when a new household comes to live in the set of municipality (i.e. in-migration). The main reasons for household elimination are out-migration and death. Three main dynamics change the household type (single, couple, with or without children and complex[5]): makeCouple; splitCouple and givingBirth. These processes are now described with more details in the same order they have been presented in this introduction.

8.1.7 BecomingAnAdult

Becoming an adult means an individual creates her own household. This can lead her to move from parental residence because of a low dwelling satisfaction level, but it's not always the case. An individual loses her child status and becomes an adult when: she finds her first job; or she is chosen by a single adult as a partner; or she remains the only children in a household after her parents leave or die while her age is higher than parameter firstAgeToBeAnAdult.

Parameter First age to become an adult—15 is the age considered by the French or other European National Statistical Offices.

8.1.8 Household Migration and Mobility

In changing residence process, we include both residential migration and mobility without making a difference, between short and long distance move, as it is often the case (Coulombel 2010, p. 66) in the literature. The submodel we propose directly manages both types of moving. However, it turned out easier for us to distinguish two categories of migration: the migration of people coming from outside to live inside the set; the migration of people who already live inside the set.

The immigration into the set is an external forcing. Each year, a number of potential immigrants from outside the set are added to the municipalities of the set.

[5] A complex household is a household which is not a single, a couple with or without children.

These potential immigrants can really become inhabitants of the set if they find a residence by themselves or by being chosen as a partner by someone already living in the set in case they are single (with or without children). Thus, looking for a place of residence is the only action they execute until they become an inhabitant of the set. Until the potential immigrant becomes a real inhabitant, she cannot search for a job. Indeed, the job occupied by people living outside the municipality set are already taken into account through the scenario and allowing potential immigrants to find a job directly would be redundant. The definition of who are potential immigrants, how numerous they are, and when they are introduced is specified exogenously. Since they are created, the potential immigrants are temporarily places into a municipality from which they can find a residence or being chosen as a partner. They are placed in a municipality following a probability to be chosen, which is computed for each municipality depending on the population size of the municipality and its distance to the frontier of the set. A particular attraction of young people for larger municipalities is also taken into account.

The mobility of people already living inside the set of municipalities is mainly endogenous. Such a mobility can lead the household simply to change residence, municipality or to quit the set of studied municipalities. Overall, a household decides to look for a new residence when:

- a new couple is formed: the couple chooses to live initially in the largest residence among the ones of the partners;
- a couple splits: one of the partners, randomly chosen, has to find out another residence even if she remains for a while in the same residence (creating her own household);
- an adult of the household finds a job away from the current place of residence (beyond the proximity parameter of the MunicipalitySet);
- a student or a retiree decides to move;
- the residence is too small or too large. This can be due to a birth, a new couple or to someone who left the residence for example. The too small or too large characteristic is assessed through a satisfaction function depending on the difference of size between the occupied size and an ideal size for this household, and the average age of the household members. In principle, people tend to move easily when they are younger and/or when the difference of size is high.

The choice of a new household is twofold: first, the household chooses a distance to move; secondly she chooses at random a new residence proposition to examine. The proposition is accepted depending on the level of satisfaction it can give. This satisfaction depends on the difference between the proposed and the ideal size, and the average age of the household members. In principle, with increasing age we assume a decrease in flexibility to accept residences different from their ideal.

A move of a household can result in increased commuting distances for some of its working members, even exceeding the proximity threshold. Such a commuter continues until she becomes the household leader through the job search mechanism and triggers the household to look for a residence closer to her job.

Parameters for Immigration Yearly migration rate; number of out of the set migrants in year $t^0 - 1$; probabilities for characteristics of the immigrants (size of the households, age of individuals…); distance to the frontier of the region of each municipality.

Parameters Within the Set of Municipalities and Out-Migration:

- The level of satisfaction of the size of the current dwelling or the one of a proposed dwelling is a function of the size of the household and of the its age composition; this function requires one parameter called β which has to be calibrated
- distribution of probabilities for an individual to accept moving over a certain distance to get a residence starting for her place of work (see (Huet 2011 p. 89) for more details)
- Laws for migration of students and retirees and acceptable distance of commuting (see for details on these processes).

Except for β, all these parameters can be extracted from the Mobility data collected in the French Census, directly or after applying some statistical tools.

8.1.9 Death

The death age of the individual is determined when she enters the simulation (through birth, initialisation or immigration). When an individual dies, its household status is updated depending on the number of remaining members and their statuses, parent or children. Households are eliminated when all their members die, when the couple splits up, or when they simply out-migrate.

Parameter Probability to die by a certain age – made available by INED from the various French Census at the national level.

8.1.10 MakeCouple

The method works as follows:

- During each time step, each single individual (with or without children) has a probability to search for a partner;
- If the individual tries to find a partner, she tries a given number of times in every municipality close to her own (her own included) to find someone who is also single and whose age is not too different (given from the average difference of ages in couples and its standard deviation); she can search among the inhabitants or the potential immigrants; the close municipalities are at a maximum distance defined by the threshold parameter "proximity" except for old people who search for a partner only in their own municipality;
- When a couple is formed, the new household chooses the larger residence (the immigrating households always go into residences of their new partners; this

move can force one member to commute very far. This situation can change only when she is becoming the leader triggered by the job search method and implying that the household will aim to move closer to her job location.

Parameters Probability to search for a partner; maximum number of trials; average difference of age of couples and its standard deviation.

The last one is given by the INSEE at the national level based on the data from Census. For the two first, they have to be calibrated since they do not correspond to existing data.

8.1.11 SplitCouple

All couples, except the potential immigrants have a probability to split up. When the split takes place, the partner who works further from the residence leaves the household and creates a new household, which implies that she searches for a new residence. When there are children, they are dispatched among the two new households at random.

Parameter Probability to split (no possible data source, has to be calibrated).

8.1.12 Giving birth

To simplify, we made the assumption that only households with a couple can have children, and one of the adults should be in age to procreate. We assumed that couple has a constant probability to have a child over the years. The parameters are the minimum and maximum ages to have a child and the average number of children by couple. From these parameters, we compute for each couple the probability to have a child during that particular year if one randomly chosen individual's age allows reproduction.

Parameters Minimum and maximum age to give birth, number of children an individual can have during her life on average. Usually ages for reproduction ranges from 18 to 45. That is the usual base to compute the total fertility rate corresponding to the number of children divided by the number of women in age to give birth during any given year. From this rate, it is possible to compute the average number of children of any simulated woman, which is about 2 for France. We can start with this value to parameterize the model. But the number of children per couple has to be calibrated since the observed fertility rate of our simulated population can vary from the value of the parameter. Indeed, the birth can only occur in couples with members having a relevant age. Consequently, the parameter number of children giving the probability of birth does not correspond with the fertility rate (which is a measure in the population, implicitly resulting from different processes leading to a birth).

8.2 Designing and Parameterising the Individual Activity

This part focuses on the design and the parameterisation of the individual activity. The purpose is to illustrate how to model in a micro simulation approach individuals' behaviour on a labour market utilising existing data. The European project that funded this work did not fund specific interviews or surveys for this purpose. But, even if such funding had been available, it would have been difficult to have a sufficiently large sample to ensure the statistical significance of the obtained attributes and behaviours. Therefore, it seemed better to use existing large database dedicating especially to the labour force, such as the labour force survey, which gives information on the labour force based on a very large sample and the weights for projection at various levels. Moreover these databases, developed by the National Statistical Office, have been built on a data collection model designed by experts. They represent common knowledge, largely shared by every stakeholder since they are used as references in decisions and predictions.

We start from existing databases and the objectives of the modelling to characterise our agents and their attributes and behaviours. That is what we discuss in the following first subsection. The two following subsections give details on the initialisation of the attributes and on the parameterisation of the behaviours. The link between attributes and behaviours is guaranteed as this data is implemented to ensure its compatibility with the agent attribute modalities. Similarly, the projection of attributes and behaviour for the whole virtual population is easy: an innovative generation population algorithm builds directly a robust and significant population of individuals while the link between modalities of attributes and their evolving rules allows an automatic projection at the population level.

8.2.1 Data Sources and Main Modelling Choices

This is to identify the agent classes and the structure of agent behaviour in each class. The first steps have been:

- to collect all relevant data source regarding the region we want to simulate considering the exact problem (aim of the project) we need to address;
- to make a state-of-the art;

From the literature and the expertise coming mainly from economists, we identify two complementary groups of dynamics to take into account to model the evolution of a local labour market:

- Job offers and corresponding dynamics;
- Job demand and occupation, and corresponding dynamics.

We identify two possible databases to help us conceptualising and parameterizing the model:

- The Census: it gives indication about the situation of individual when being student, retired, or active and also who is occupied and who is not occupied, what occupations individual have aggregated in socio-professional categories and activity sectors; Census data are available at the municipality level for three different dates 1990, 1999 and 2006. We can also benefit from the mobility tables of the Census giving, at least in 1999, an exhaustive description of the commuting flows between municipalities; French Census data are also available for 1982 but not electronically;
- Labour force survey (from 1990) and census data;

From literature and data, we have to define agents:

- corresponding to the local level of offer: the **municipality**
- corresponding to the job demand and occupation: the **individual** is the one who is going to search for a job, deciding if and where she searches taking into account the **household** of which she is a member and her *municipality* of residence.

Then we have a municipality offering jobs, composed from households, themselves composed of individuals who decide, considering their household, if and where they are going to search for a job. A job can be found in a municipality and individuals accept found jobs based on the distance.

Other available data sources include SIRENE and UNEDIC. The SIRENE database includes information on the number of societies by activity sector. The UNEDIC database includes the number of paid employees by activity sector. But both these data sources describe only a part of our problem and start only in 2000 while the simulation requires longer periods to allow for a proper calibration of the model.

The incompatible coverage also constrains the choice of agents and their attributes. However, given the available datasets we decide to start simulations in 1990. On the one hand, it means some the parameterisation of some attributes is less robust than with shorter calibration periods. A later start would allow us to use the supplementary information given in more recent surveys and not available in older surveys. For example, we use only four modalities of size to describe the size of dwellings because only four are available in 1990 while five and more are recorded in later surveys. On the other hand, the 1990 census data give us the cross distribution socio-professional categories x sector of activities we use to define the jobs while this cross distribution is not available later. Then, we can and have to use IPF to define the job offer after 1990 starting from the 1990 cross distribution.

The definition of a job is directly driven by the available data. Both Censuses and Labour Force Survey (or Employment survey) describe jobs with profession (socio-professional category) and activity sector. Both also contain data on age and situation (student, retired, actives, occupied or not, inactive) allowing us to make a connection between both sources of data. Moreover, when the data sources are "official", it often corresponds to the common knowledge of stakeholders and other decision makers.

Moreover, as a general modelling good practice, it is particularly important to minimise the number of unknown parameters. Indeed, every parameter which is not derived from the data has to be calibrated. The calibration computational cost increases with the number of parameters. Moreover, the more numerous are the parameters to calibrate, the less relevant also is likely to be the model which, given its large number of freedom degrees, can produce almost any trajectory.

8.2.2 Defining the Initial Individual Labour Attributes

The main source of information to define attributes and their values is Census data. The French Census is available for 1990, 1999 and 2006. The 2006 Census has to be used with caution since it is different from 1990 and 1999. It is now a continuous survey which interviews a part of the population every year. Municipalities having less than 10,000 inhabitants are exhaustively surveyed by 1/5 every year. Larger municipalities are sample surveyed every year. In both cases, INSEE, responsible for the Census, give the information allowing the projection at the population level every year. A very good point is that the access to data is easy and free.[6]

To compute a population with sufficiently realistic local statistical properties for individuals and households, we propose an algorithm described in (Gargiulo et al. 2010, p. 7) presenting the generation of households in the Auvergne Region. An improved version has been developed for generating the Cantal population. To summarize our algorithm, we build for each municipality a list of agents with the exact number of individuals being each age and a list of households with the exact number of household members. Then, we try to fill one by one each household with individuals taking into account the probability of households having some particular properties, such as being a couple or having a given number of children. Each time a household is completed, another one is selected to be filled. At the end, we have a virtual population of households following the exact distribution of sizes, having good statistical household properties and composed from individuals following the exact distribution of ages. To built the initial population of Cantal, our algorithm uses for each municipality:

- The distribution of the size of households—available at the municipality level in 1990.
- The distribution of ages of individuals—available at the municipality level in 1990.
- The distribution of ages of the reference person of households—available at the municipality level in 1990.
- The distribution of household types (single, couple, couple with children, single-parent, other)—available at the municipality level in 1990.

[6] made available by the Maurice Halbwachs Center of the Quételet Network (http://www.reseau-quetelet.cnrs.fr/spip) for 1990. For 1999 and 2006, they are directly accessible through internet via the website of INSEE http://www.recensement-1999.insee.fr/ and http://www.insee.fr/fr/publics/default.asp?page=communication/recensement/particuliers/diffusion_resultats.htm).

- The distribution of age differences for couples—only available at the national level in 1990.
- The distribution of the probability to be a child (i.e. living at parental home) by age and for each household type—available at the municipality level in 1990.

This generation method is different from the nowadays used IPF (Iterative Proportional Fitting) which reweight a measured population under some constraints to obtain a virtual population representing the one the modeller is interested in. However this method cannot control the attributes at both levels, the person and the household. Some recent work proposed a hierarchical IPF (Müller and Axhausen 2011, p. 91) to control the two levels but they still required an initial sample, which can be reweighted to fit the scale the model is interesting in.

After the virtual population has been built, individuals require a labour market status. That means the following four individual attributes have to be parameterised during the initialisation: Activity status; Profession, approximated by the socio-professional category; Sector of activity to define, with the profession, the occupied job; Place of work.

To characterize the status we distinguish between active and inactive individuals. Active people can be employed or unemployed. For non-active people we distinguish three categories: students, retired and other. No further characterization is required for non-active person. On the contrary, active people, both employed and unemployed require a socio-professional category (SPC) defining their profession. Moreover, employed individuals require a sector of activity defining the occupation (see Sects. 81.1.3 and 8.1.1.5 for details). Once the municipality of employment is determined, the employed individual is successfully parameterized.

Figure 8.2 shows the generation algorithm. The initialization of the activities starts from the population of households previously generated for each village: each person is assigned an activity, according to the characterization presented above. All the individuals younger than 15 are automatically considered students. For all the others the first step is the decision about being active or not. This decision depends on the age of the person. If the person is not active then her age determines whether she is retired or a student. If she is neither student nor retired, she will be identified with the status "inactive". If the person is active, the first step is the selection of the socio-professional category (SPC). This choice depends on the age. Secondly it is decided whether the person is employed or unemployed, according to the age. If she is unemployed, no further choices are needed. If she is employed, the municipality of employment is determined. The municipality of employment depends on two questions: first, does she work inside her municipality of residence? If no, find at random a place of work among the possible places of work starting with her own municipality of residence if employment is available according to the SPC. The possible places of work are defined through a generated virtual network built from the mobility data of the French Census of 1999 (see the generation model proposed in (Gargiulo et al. 2011, p. 69) and improved in (Lenormand et al. 2012b, p. 68)). Finding a possible place means the individual can find a free job partly defined by the same SPC as hers. A vector for available jobs is maintained (corresponding to the total number of commuters-in at the beginning of the initialisation) for each municipality

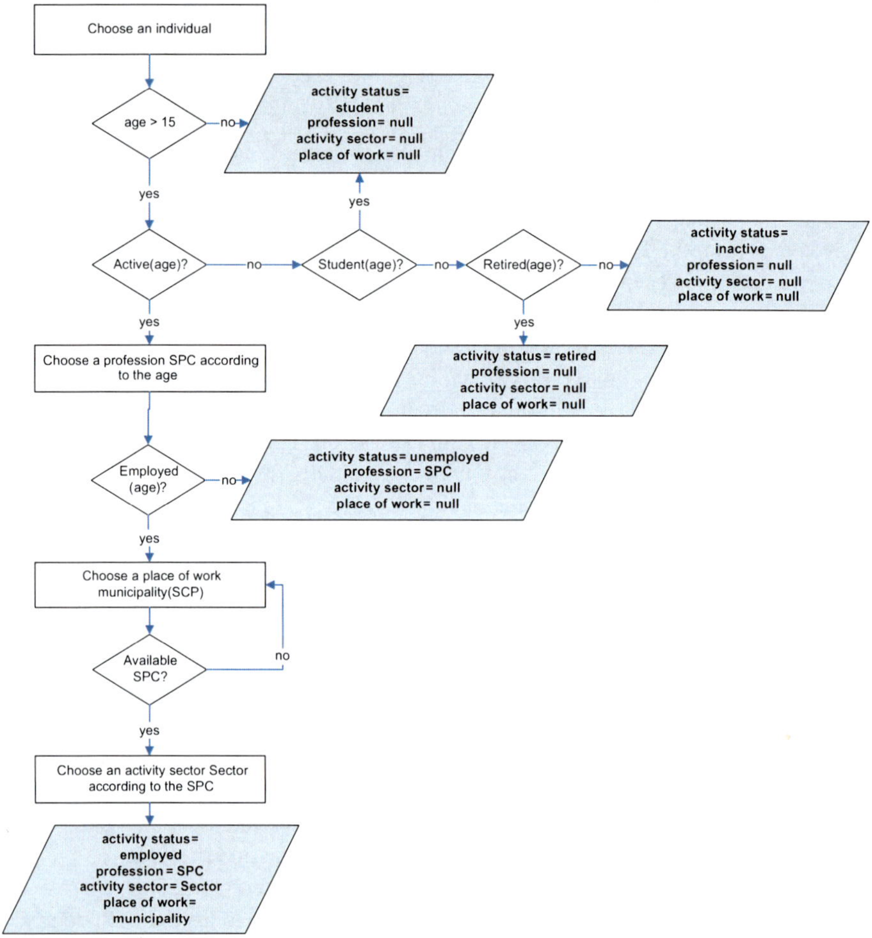

Figure 8.2 Algorithm for the initialization of the activities for Auvergne case study

and decreases with individuals filling vacancies. If no vacancies remain among the possible places of work while an individual is still looking for employment, the attribution of a place of work among the possible ones is forced. Indeed, this can occur due to the fact the generated virtual network is built under the only constraints related to the job demands and the job offers of each municipality. The virtual network doesn't consider the SPC then it can't ensure a demand with a particular SPC can be satisfied by an offer with this SPC in the set of municipalities it has fixed as possible places of work. Finally, an activity sector is attributed to the employed individual based on the cross distribution SPC. We have to acknowledge that the French Statistical Office, as many Statistical Offices, use two ways to count the jobs: counted on the place of residence—that means corresponding to the job occupation by people living in a municipality wherever they work; and counted on the place of work—that means counted on the municipality where people work wherever they live. The algorithm uses the following data for each municipality of the set:

- Age x activity status counted on the place of residence
- Age x SPC for actives counted on the place of residence
- Distribution of probabilities working inside her place of residence by SPC
- A generated commuting network through (Gargiulo et al. 2011 p. 69) (Lenormand et al. 2012b, p. 68) given for each municipality the distribution of commuters out to each of the other municipality
- SPC for actives x activity sector counted on the place of work

8.2.3 Defining the Individual Behavioral Rules Regarding Activity

This part is dedicated to the parameterisation of events on the labour market. Characterization and parameterization is required for those rules that change the value of the individual's attributes related to its labour activity: Activity status; Profession, approximated by the socio-professional category; Sector of activity to define, with the profession, the occupied job; Place of work.

The main data source to do so is the European Labour Force Survey, and particularly its French declination called in French "Enquête Emploi", meaning "Employment survey". The data are kindly made available for free by the Maurice Halbwachs Center of the Quételet Network.[7] This Employment survey was launched in 1950. It was redesigned in 1968, 1975, 1982, 1990 and 2003. From 1982, the survey became an annual survey. Since the last redesign the survey is implemented continuously to provide quarterly results. The resident population comprises persons living on French metropolitan territory. The household concept used is that of the 'dwelling household': a household means all persons living in the same dwelling. It may consist of a single person, or of two families living in the same dwelling.

As our approach starts the simulation in 1990 the first period is based on annual data while from 2003 on values can be considered in quarterly time steps (Goux 2003 p. 56; Givord 2003; the data to select from these two periods vary a bit due to the structural and practical changes in the survey).

Coming back to the description of the whole data, the sample sizes of the data varies from 168,883 to 187,326 from 1990 to 2002 each year and from 92,300 to 95,647 each quarter a year for the new Employment survey. The individuals are asked a very comprehensive series of questions from 1990 to 2006, related to their work. In particular, we can follow their situation year by year, and also their wishes to change job and the type of job they are looking for. Table 8.4 shows the variables we extract from the databases to compute the probabilities we need. However, for the sake of simplicity, we use only data from 1990 to 2002 to explain how to extract the information we need from the data.

[7] http://www.reseau-quetelet.cnrs.fr/spip/.

Table 8.4 Data to extract from the various databases of the French labour force survey to compute the probabilities related to working status of the individual

1990–2002	2003	2004	2005	2006	2007	Meaning of the variable
ag	Ag	Ag	ag	Ag	Ag	Age
annee	annee	Annee	annee	annee	annee	Year of interview
dcse	csepr	Csepr	csepr	csepr	csepr	Socio-professional category
cspp	cspp	Cspp	cspp	cspp	cspp	Socio-professional category of the father
dcsep	cser	Cser	cser	cser	Cser	Socio-professional category one year before
dcsea	cslong	Cslong	cslongr	cslongr	cslong	Socio-professional category which has been occupied for most of the time (for inactive and unemployed people)
tu99	tu99	tu99	tu99	tu99	tu99	Urban area type
fip	eoccua	Eoccua	eoccua	eoccua	eoccua	Occupation one year before
extri	extriA, extriA04	extri99 extri04, extri05,	extri05, extri06,	extri06	extri06	Weights making the interviewed individuals representative (depending on the census done 1999 or of the first result from the last French census (in 2004, 2005, 2006)
rg	reg	Reg	reg	reg	Reg	Region of residence
fi	sp00	sp00	sp00	sp00	sp00	Occupation during the month of interview
–	trim	Trim	trim	trim	trim	For the second period of the survey, the only keep the first quarter of the year
csrech	csrech					Searched socio-professional category
dre1						Situation in regards to employment (mainly to use dre1 = 5 meaning people looks for a job (or another job))
	soua; mrec					Wish another job; Is the individual has searched for a job during the last four weeks?

From the databases, we considered only the population being more than 14 that is not military people of students (FI = 3 and 4)

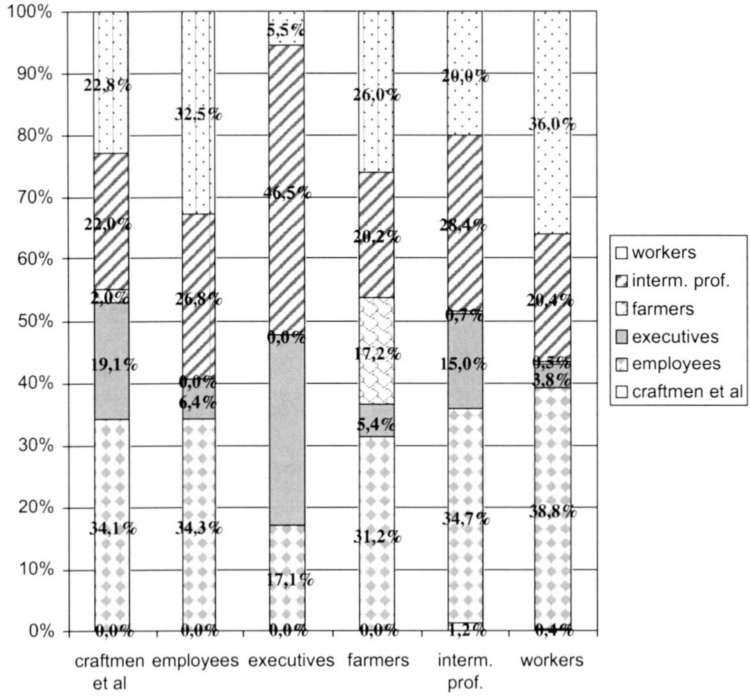

Figure 8.3 Distribution of SPCs choices by children regarding the father's SPC (in abscissa) for the Auvergne population. (Source: French Labour Force Survey 1990–2002 data)

8.2.3.1 Entering the Labour Market

A first step consists of extracting the age from which on the individual is going to look for a job. This will determine the age at which a student status changes to a "on labour market" status. We consider in the period 1990 to 2002 the value FIP = 3, which means that the individual was student the year before and the value FI = all the possible values except 3 means that the individual is not a student anymore. Then, for each five-year step we compute the probability to be a given age and having entered on the labour market for every year.

We used the weights to obtain a projection of the data at the Auvergne level. Auvergne is the region containing the Cantal "département" and three others. That is the closer significant and representative level of the Cantal. Then, we assume the probabilities are the same at the regional and the "département" level.

The second step is to allocate a first SPC (proxy used for defining the profession) to the individual allowing us to approximate what she is going to look for. We know that both these variables, the age of entry and the first SPC, are not independent. Moreover, a social determinism rules the choice of the profession by children compared to the profession of their parents. Figure 8.3 presents such a relation for the Auvergne population. It shows, for example, that almost only farmers' children become farmers or that executives' children mainly become executives and/or adopt an intermediary profession.

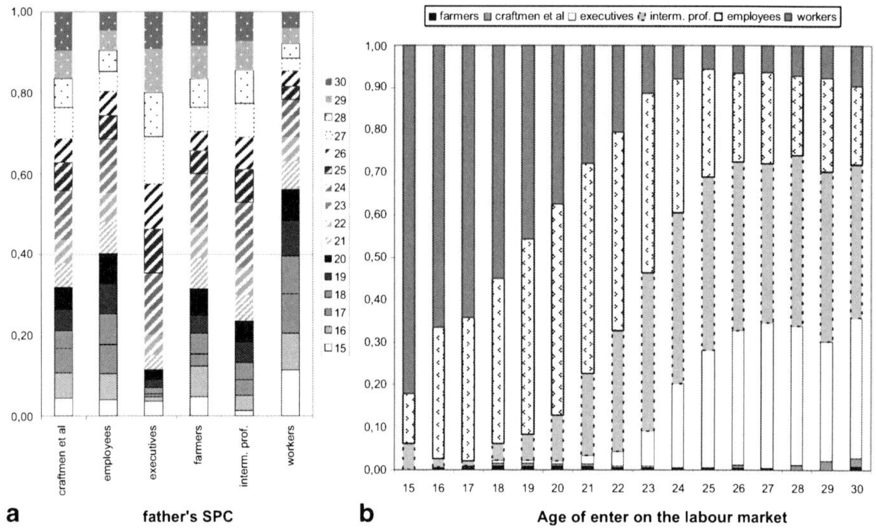

Figure 8.4 (**a** on the *left*) Probability of a "first" SPC depending on the age of entry in the labour market; (**b** on the *right*) Distribution of probability to enter the labour market at a given child age for each of the six father's SPC considered—French population. (Source: French Labour Force Survey, 1990–2002 data)

Thus, starting from this social determinism, we have some indications to set the SPC of children. However, we also have to decide the age of entry in the labor market, and we know that this age is not independent from the level of education, which can be related to the SPC. Consequently, we apply a two-time process which, at first, decides the age at which to enter the labor market using the father's SPC and then determines the child's SPC depending on the age of entry.

The age of entry on the labour market is determined by the SPC of the father. Since the individual has no gender in our model, the father is randomly chosen between the two parents when there are two.

A criticism can be formulated to this approach since the SPCs of the couple members is not controlled, while we know from the literature that the partner is not chosen at random regarding her SPC (Bozon and Héran 1987, p. 50). The homogamy can be explained by the constraint associated to the meeting places (Bozon and Héran 1988, p. 51). It has been identified as a possible next step for modelling.

Figure 8.4a shows the distributions of probabilities to enter the labour market depending on the various ages of a child for each of the six SPC attributed to the father. We can for example read that if the father is an executive, the probability to enter on the labour market before 20 is only 0.1 while it is more than 0.5 if the father is a worker. Once our individual has an age to enter the labour market, we can determine her first SPC. Figure 8.4b shows for each age of entry on the labour market (abscissa) the distribution of probabilities over the possible SPC to provide the individual with a first SPC. For example, one can notice how high the likelihood of looking for a worker position for the individual looking at first for a job at

15 is, while at 30, she will mostly look for intermediary or executive positions. The individual who enters the labour market can decide looking for a job.

8.2.3.2 Individual Job Searching Decision

We assume that the probabilities are stable in time for the Auvergne region. Thus, we mix the data from the years 1990 to 2007 in a single sample. Starting from the variables presented in the Table 8.4, we count the frequencies of transitions between inactive, unemployed, employed, from 1 year to the following. For each counted transition, we take into account the weight of the related individual in order to have a probability quantified for the Auvergne level.

Finally, we calculate the probability to reach a given situation by dividing the total obtained for a transition starting from the situation x by the sum of all the totals related to the transitions starting from this same situation x.

We focus on the municipalities of the Auvergne region having less than 50,000 inhabitants using the area type "tu99".

8.2.4 From and to the Inactive Status

The following variables are used to extract the transitions from a starting situation to an arriving situation. They are used for the transitions from and to the inactive status.

- fip=7 plus 8 or EOCCUA=6 plus 7 to define the inactive status as starting situation; fi=7 or SP=8 to define the inactive status as arriving situation;
- fip=2 or EOCCUA=2 to define the unemployed status as starting situation; fi=2 or sp00=4 to define unemployed status as an arriving situation;
- fi=1 or EOCCUA=1 to define employed status as starting situation;
- DCSP or DCSA are used to define to starting SCP for unemployed and employed while DCSE is used to define the arrival SCP (for unemployed).

The Table 8.5 shows the extracted probabilities for the Auvergne region.

8.2.5 Probability to Look for a Job with a Given Profession

The probabilities are computed using the same method we used to compute the probabilities of transitions of activity status. The difference is that we use the answers to the questions about the fact that the interviewee looks for another job. For the first period, we select the employed individuals (fi=1) looking for a job (drel=5). For the second period of the survey, from 2003 to 2007, we assume people look for a job if they have answered SOUA=1 (want to have another job) and MREC=1 (have searched for recently) or SOUA=1 and MREC=2 and NTCH=1 or 2 (have not recently search for because they wait for answer to recent applications or they have been ill for a while).

Table 8.5 Probabilities of the transitions "inactive → unemployed", "unemployed → inactive depending on SPC", "employed → inactive depending on SPC"

		Arriving situation	Inactives	Unemployed					
Starting situation	**Starting SCP**	**Arriving SCP**		farmers	craftmen et al	executives	interm. profes.	employees	workers
Inactives				0,00005557	0,00055947	0,00031037	0,00172877	0,00644310	0,00604629
Unemployed	farmers		0,05462738						
	craftmen et al		0,06335331						
	executives		0,11808481						
	interm. profes.		0,06202433						
	employees		0,07066007						
	workers		0,06165634						
Employed	farmers		0,00650018						
	craftmen et al		0,01423226						
	executives		0,01729000						
	interm. profes.		0,01192824						
	employees		0,00930251						
	workers		0,01129013						

Table 8.6 Probability for unemployed people to search for a job with various SPCs knowing the current SPC of the individual

SPC/looks for	Farmers	Craftsmen et al.	Executives	Interm. prof.	Employees	Workers
Farmers	0.000	0.000	0.000	0.177	0.376	0.447
Craftsmen et al.	0.000	0.079	0.012	0.088	0.443	0.377
Executives	0.000	0.037	0.499	0.256	0.171	0.037
Interm. prof.	0.000	0.009	0.053	0.591	0.273	0.074
Employees	0.003	0.007	0.006	0.063	0.808	0.113
Workers	0.006	0.010	0.003	0.056	0.251	0.674

8.2.6 Deciding Looking for a Job When Unemployed

Unemployed people are assumed to be those who search for a job. Even if, in the labour force survey, only 80 % of unemployed people declare searching a job, we assume the probability to search for a job of unemployed people is one. Indeed, if we consider the whole model, it globally underestimates the job offer and the probability to find a job. This is difficult to correct as, for instance, we cannot consider that in most cases a job offer is proposed before it has been quit while the model time step is not less than 1 year. Also we assume the job offer equal to the job occupation. Then, the probability to search for a job of unemployed people is one in order to compensate a bit this underestimation and be able to occupy every job offer (which is the state the model has to reach). The data indicates the probability to look for a job for unemployed individuals is quite stable until 54 years of age and dramatically decreases for older individuals. A second step of the modelling work would be to see if this dramatic decrease needs to be considered. We also analyse how different parameters describing the household (the number of unemployed in the household, the number of children, or the type of household) influence the probability to look for a job, and we did not find any clear dependency.

The probability to begin searching (i.e. becoming unemployed) if an individual did not search previously (not because she is employed) corresponds in the model to the transition from inactive to unemployed. As already mentioned, it is the complementary value for each age range of the value to make the transition from inactive to inactive.

Since an individual is unemployed, it is necessary to define which SPC she is going to search for. It varies a lot with the current SPC of the individual. As shown in Table 8.6 even if there is a tendency to look preferentially for her own SPC, an unemployed individual can prefer changing SPC. That is particularly the case of farmers and craftsmen. Then, we parameterise the process from the computation of the probability distribution to choose a SPC knowing the current SPC.

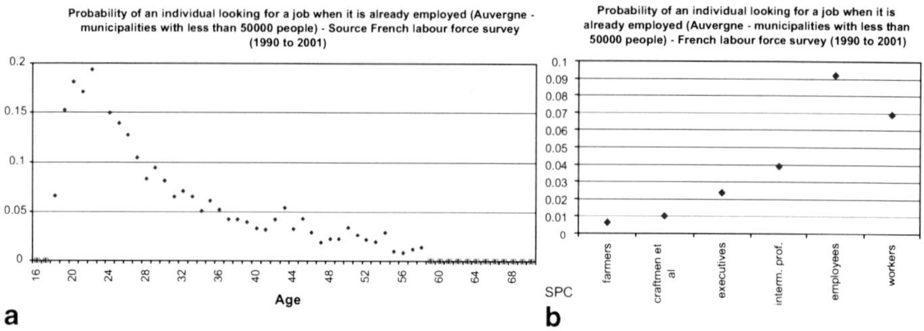

Figure 8.5 **a** Probability for an already employed individual to look for another job according to the age (on the *left*); **b** Probability that an already employed individual looks for another job according to socio-professional category (on the *right*)

8.2.7 Deciding Looking for a Job When Already Employed

We consider those respondents being employed who answered that they are looking for another job. We have the age of these people, as well as the type of their current job. The analysis shows that the age is a very significant variable for determining if an employed individual looks for another job (see Fig. 8.5a). Young people are more susceptible to look for another job and this tendency decreases with age.

The SPC is also a significant variable to predict the probability to look for a job (see Fig. 8.5b). Some SPC, such as employed farmers or craftsmen are not very susceptible to look for another job. On the contrary, others, such as workers and especially employees have quite a high probability to look for another activity.

Table 8.7 shows the parameter values for the decision searching for a given profession when the individual is already employed for some age ranges. For employed people, we built a probability containing the both information have decide to search for a job and what she searches for. It is important to point out that the probabilities presented in Table 8.7 do not add up to one but to the overall probability to search, which is quite low for already employed people.

8.2.7.1 Individual Searches for a Job

Since the individual knows which profession she wants to search for, she has to find a place where to look for a job. Firstly, the individual selects an accepted distance she would want to commute. The next section presents how to the related probabilities. If the chosen distance is higher than zero, the individual has to decide if she is going to work outside her set of municipalities. The law allowing this decision and the way to extract it from data is the subject of what follows in the next section. In case the individual has not found a job, she revises the maximum distance. She revises the distance up to 10 times.

Table 8.7 Extract of probabilities for employed people with a given SPC and a given 5-year old age to look for a job within a given SPC

Age range	Looks for/is a	Farmers	Craftmen et al.	Executives	Interm. prof.	Employees	Workers
15	Farmers	0.0000	0.0000	0.0000	0.0000	0.0000	0.0002
	Craftmen et al.	0.0000	0.0000	0.0000	0.0000	0.0011	0.0014
	Executives	0.0000	0.0000	0.0000	0.0000	0.0010	0.0000
	Interm. prof.	0.0000	0.0000	0.0000	0.0000	0.0143	0.0040
	Employees	0.0000	0.0000	0.0000	0.0000	0.1319	0.0168
	Workers	0.0000	0.0000	0.0000	0.0000	0.0162	0.0498
…	Farmers	…	…	…	…	…	…
	Craftmen et al.	…	…	…	…	…	…
	Executives	…	…	…	…	…	…
	Interm. prof.	…	…	…	…	…	…
	Employees	…	…	…	…	…	…
	Workers	…	…	…	…	…	…
55	Farmers	0.0000	0.0000	0.0000	0.0000	0.0000	0.0000
	Craftmen et al.	0.0000	0.0000	0.0000	0.0000	0.0002	0.0002
	Executives	0.0000	0.0000	0.0000	0.0000	0.0002	0.0000
	Interm. prof.	0.0000	0.0000	0.0000	0.0000	0.0030	0.0005
	Employees	0.0000	0.0000	0.0000	0.0000	0.0274	0.0021
	Workers	0.0000	0.0000	0.0000	0.0000	0.0034	0.0062

8.2.8 The Probability to Accept a Distance to Cross Over to Work

The distance of search for a job is selected from a probability law giving the probability to accept a certain distance between the residence and the work place. The principle is very simple: the probability to commute at a given distance i [$pc(i)$] is assumed to be the product of a probability to accept a certain distance i [$pa(i)$] by the pay offered at i [Oi] with a renormalisation coefficient k: $pc(i) = k\ pa(i) * Oi$.

Then, it is possible to extract the probability to accept a given distance (pa) to work place, which will be used in the model. This procedure, coupled to an appropriate job offer, will allow maintaining the statistical properties of the pc distribution over the time of the simulation.

We extract from the mobility data of the 1999 Census for every municipality of the Auvergne region data on commuting (pc) and data on job occupations, which we assume to be equivalent to job offers (O). Evidently, the number of occupied jobs is used as a relevant proxy for the job offer of a municipality. An exhaustive description of the work allowing to build this probability law is given in (Felemou 2011, p. 76; Fig 8.6).

Figure 8.6 shows an example of commuting data probability distribution (DDC = pc) and of job offer probability distribution (DOE = O) for one randomly chosen municipality

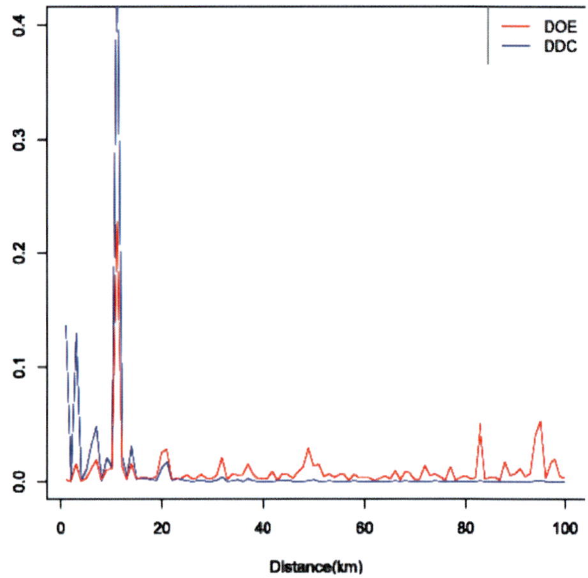

Figure 8.6 Example for one municipality of the density distribution of job offers (DOE = O) and the one of commuters (DDC = pc)

A classification of acceptable distance distributions shows municipalities can be classified in three different groups, apparently depending on the size of the municipality of residence (see Fig. 8.7 on the right). Thus, we assume for this parameter three probability distributions shown on the left of Fig. 8.7 for three different size-dependent classes of municipalities (to the right of Fig. 8.7). The data suggests that the larger the municipality, the lower the probability to work in the place of residence and the longer the commuting distance.

It is important to emphasise that only if the selected distance is higher than zero, the individual has to decide if she is going to outside or inside the set.

8.2.9 *Going to Work Outside the Set*

When the individual is commuting—meaning she has picked out a distance of research higher than 0—she has to check if she has a chance to commute outside considering her place of residence. Indeed, an individual living close to the border of the set has a higher probability to commute outside the set. Then, the individual chooses at random to work outside depending on the probability associated with her municipality of residence. Each municipality has such a probability which is a function of its distance to the border of the set. This function is extracted from the mobility data from 1999 (Source: INSEE 1999). Figure 8.8 shows this function for the Cantal department and the whole Auvergne region of which Cantal is a part. Both laws are quite close and it appears relevant to use as a parameter the law extracted for the whole region since it is probably less noisy.

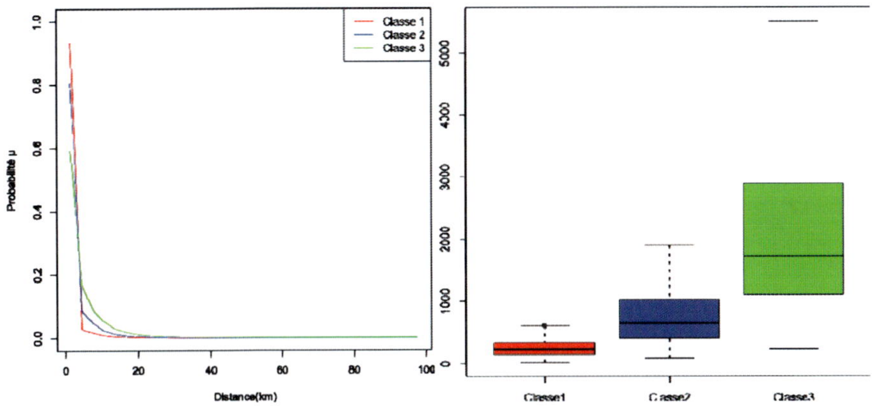

Figure 8.7 Probability laws that an individual accept to a certain commuting distance knowing that a job is available for it.(on the *left*)—different population sizes for the municipalities of each sub-group (on the *right*)

Figure 8.8 Probability to commute outside the set (ordinate) depending on the distance of the municipality of residence to the frontier of the set (abscissa in Euclidian kilometers)—*Red* Cantal, *Blue* Auvergne

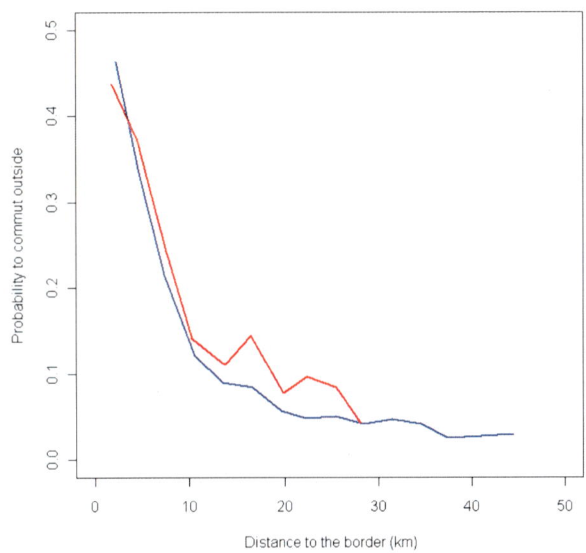

We are now describing how to extract the probability law for the final event which is going on retirement.

8.2.9.1 Going on Retirement, and Stop Searching for a Job

To extract the transition to the retirement, we consider, in the period 1990–2002, the value FIP = all except 5 or 6, which means that the individual has not yet retired and

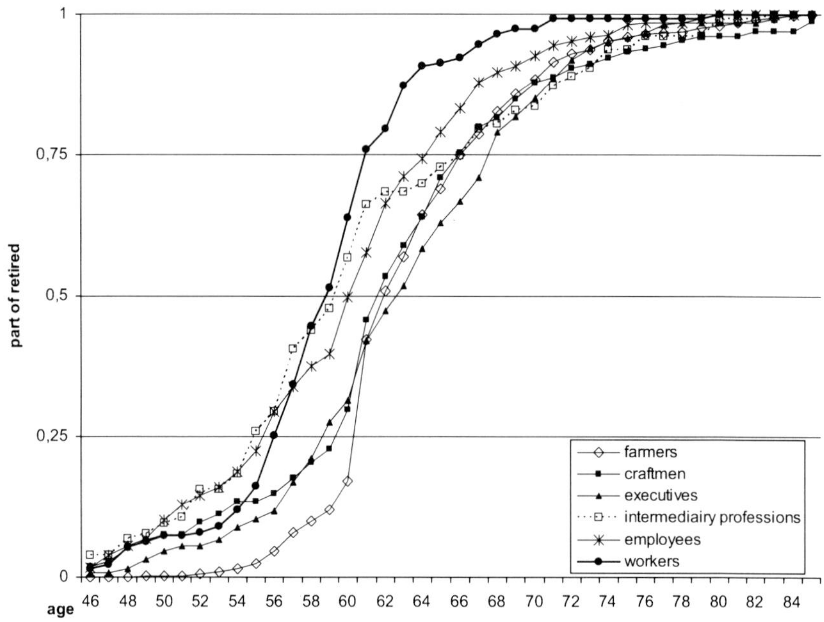

Figure 8.9 Speed of going into retirement by SPC (source LFS)—France level

the value FI = 5 or 6, which means that the individual is now retired. We assume that the retiree does not search for a job anymore since this is generally the case true in France. Figure 8.9 shows that the speed of transitioning into retirement varies a lot from one SPC to another: we can read for example that at 60, 63 % of workers are retired while only 17 % of farmers are retired. Then, instead of considering a generic retirement law for all the individuals we consider a law for each SPC. Indeed, as these laws influence the job availability at a given moment it is very important to be sufficiently precise.

8.3 Lessons/Experience

First, we want to stress the necessity to not only consider the objectives of the model during the design, but from the very beginning exploring existing data sources and studying the implicit model beside the existing databases. The availability of data and the more or less implicit model guiding the collection of data constrain the definition of agents, their attributes and behaviours.

Using large existing databases can appear more relevant, especially the "official" ones from the National Statistical office, than collecting a small sample and reweighting it to obtain a statistically significant artificial population.

For these large databases, the models guiding the collection of data represent the expertise knowledge and generally assume some dynamics, particularly if time series are collecting during the survey. Moreover, if the data sources are collected by the National Statistical Office, they probably represent the commonly used information and knowledge by the stakeholders and policy makers. A model which aims to inform decision making is more useful if it can be easily understood and discussed by the relevant decision makers. This is easier if the model starts with common knowledge.

More generally, the modeller has to identify the rationale behind the considered data sources and use it to build the dynamic model. Indeed, this rationale often makes some implicit assumptions on the dynamics. Let's take the definition of a household as an example. "*In surveys prior to 2005, people were required to share the same main residence to be considered as households. It was not necessary for them to share a common budget. De facto, a household corresponded to a dwelling (main residence)*". Thus, until 2005, the French National Statistical Office (INSEE) assumes the household/family is defined by the place where it lives, which is unique. Indeed, following the INSEE definition, each person in a household may belong to only one family. In this framework, residential mobility is a household/family decision and the number of occupied dwellings in a place corresponds to the number of resident households. That is also what we assume in the model. "*Since 2005, a dwelling can include several households, referred to as "living units". Every household is composed of the people who share the same budget, that is who contribute resources towards the expenses made for the life of the household; and/or who merely benefit from those expenses.*" The new definition is based on the fact that related or unrelated individuals can share the same budget and have a habitual residence (the dwelling in which they usually live). This new definition takes into account some cultural evolutions and allows a European homogenization of the way households are defined. However, it modifies the way the dynamic of move can be considered since each individual of the household can have more than one dwelling. This is to point out that the choice between one data source and another corresponds to a representation of the world to which some particular dynamics can be linked. If the first definition of household is more related to the idea that relationships between people can be identified by the concept of family and/or the identical of place of living, the second definition puts the economical constraints (i.e. the sharing budget) much more at the heart of the dynamics of closeness. A modeler, having the choice between a data source containing data built on the first definition and another one based on the second definition, should be aware of the choice to make and communicate about it. Thus, choosing to only use data on the SCP and the activity sector to describe a job while it is possible to use the salary, which is available in some databases, makes having an occupation much more important than the level of salary. It also implies, for example, that an individual can change jobs just to change their working environment. Differently, the classical economic models considering job change start from the salary and assume an individual changes to increase their salary. We simply assume our individual wants to change jobs, without necessarily changing SCP at the same time. However, one can notice our assumption is relevant due to the existence of a minimum salary in France which ensures a minimum amount of money to live with.

The choice of existing databases for facilitating model design and parameterisation needs to consider:

- a longer as possible period of calibration: indeed it is not sufficient to strongly link the model to data if the model is not calibrated or calibrated with poor data compromising the robustness of the trajectory of underlying model dynamics;
- a sufficient number of modalities for each attribute in order to be able to reproduce the diversity of relevant agent types and behaviours. For example, we chose to aggregate in our work jobs in 24 types; at the end this depends on data availability;
- a minimum number of variables to calibrate: too many unknown parameters implies we don't know much about the dynamics and every explanation for observed trajectories can be valuable;
- the possibility to use them simultaneously for initialising agent attributes and defining agent behaviours: that means in particular that they have to have common variables allowing for a link between them. The challenge is to make an easy fit between attributes and behaviours.

Finally, starting from large national databases makes it likely that the model can be easily implemented and parameterized in another country. For instance, the example on the individual dynamics of activities indicated the possibility to apply the model in another European country even if some small adaptations are required. Indeed, Europe tends to harmonise the data bases in order to have common indicators at the European level. Then, large national databases have been designed or redesigned for answering the European demand. For example, the French "Employment survey" is the data source for the French contribution to the European Labour Force Survey. That is why (Baqueiro Espinosa et al. 2011, p. 83) proposes a way to parameterise our model directly starting from the data of this European survey. For the same reason, national census data in Europe tend to consider more and more comparable or identical variables. That makes it possible to use them to parameterise our model even if a particular attention to the definition of used concepts remains: while to be a retiree in France (at least until a very recent period) means not looking for a job, it is not the case in UK for example.

Taking into account data at an early stage is not an easy task. It is at the same time laborious and confusing since the modeller is confronted with a very large set of information and more or less implicit knowledge. Finding a way to use the data and to choose the object, their attribute and the dynamics in order to remain simple as possible is much more demanding than developing a theoretical model. However, for such complex systems and models as ours that focus on the dynamics of interacting municipalities, the approach allows to properly define and control some sub-dynamics, even if they are not independent from other dynamics in order to test hypothesised system properties. For our concerns, we expect the expertise we developed for the labour market in conjunction with the robust parameterisation of the individual activity dynamics and job offer dynamics, will allow us to better understand how the demography impacts on the population/depopulation phenomena and how these phenomena impact on demography in return.

Acknowledgements This work has been funded under the PRIMA (Prototypical policy impacts on multifunctional activities in rural municipalities) collaborative project, EU 7th Framework Programme (ENV 2007-1), contract no. 212345.

References

Aubert F, Dissart JC, Lépicier D (2009) Facteurs de localisation de l'emploi résidentiel en France, XLVIème Colloque de l'Association de Science Régionale de Langue Française (ASRDLF), 6–8 juillet, Clermont-Ferrand, France, 27

Ballas D, Clarke GP, Wiemers E (2005) Building a dynamic spatial microsimulation model for Ireland. Popul Space Place 11:157–172

Ballas D, Clarke GP, Wiemers E (2006) Spatial microsimulation for rural policy analysis in Ireland: the implications of CAP reforms for the national spatial strategy. J Rur Stud 22:367–378

Ballas D et al (2007) Using SimBritain to model the geographical impact of national government policies. Geogr Anal 39:44–77

Baqueiro Espinosa O et al (2011) Two adaptations of a micro-simulation model to study the impacts of policies at the municipality level. In PRIMA European project (ed) Working paper (p 62). Newcastle University, Cemagref

Berger T, Schreinemachers P (2006) Creating agents and landscapes for multiagent systems from random sample. Ecol Soc 11(2):19

Birkin M, Clarke M (2011) Spatial microsimulation models: a review and a glimpse into the future. In J Stillwell and M Clarke (eds) Population dynamics and projection methods (Understanding population trends and processes Vol 4, pp 193–208). Springer, Netherlands

Birkin M, Wu B (2012) A review of microsimulation and hybrid agent-based approaches. In AJ Heppenstall et al (eds) Agent-based models of geographical systems (pp 51–68). Springer: Netherlands

Blanc M, Schmitt B (2007) Orientation économique et croissance locale de l'emploi dans les bassins de vie des bourgs et petites villes. Economie et Statistique 402:57–74

Bousquet F, Le Page C (2004) Multi-agent simulations and ecosystem management: a review. Ecol Model 176:313–332

Bozon M, Héran F (1987) La découverte du conjoint: I. evolution et morphologie des scènes de rencontre. Population 42(6):943–985

Bozon M, Héran F (1988) La découverte du conjoint: II. Les scènes de rencontre dans l'espace social. Population 43(1):121–150

Brown DG, Robinson DT (2006) effects of heterogeneity in residential preferences on an agent-based model of urban sprawl. Ecol Soc 11(1):46

Brown DG, Aspinal R, Bennett DA (2006) Landscape models and explanation in landscape ecology—a space for generative landscape science. Prof Geo 58(4):369–382

Champetier Y (2000) The (re)population of rural areas. Leader Magazine 23(Springer Special Issue)

Coulombel N (2010) Residential Choice and Household Behavior: State of the Art, (Working Paper 2.2a: Ecole Normale Supérieure de Cachan), 69.

Davezies L (2009) L'économie locale "résidentielle". Géographie économie société 11(1):47–53

Deffuant G. et al (2001), Rapport final du projet FAIR 3 2092 IMAGES: Modélisation de la diffusion de l'adoption de mesures agri-environnementales par les agriculteurs (1997–2001).

Deffuant G et al (2002) How can extremism prevail? A study based on the relative agreement interaction model. JASSS 5(4):4

Deffuant G, Huet S, Amblard F (2005) An individual-based model of innovation diffusion mixing social value and individual payoff dynamics. Am J Soc 110(4):1041–1069

Deffuant G, Skerrat S, Huet S (2008) An agent based model of agri-environmental measure diffusion: what for? In A Lopez Paredes, C Hernandez Iglesias (eds) Agent based modelling in natural resource management (pp 55–73). INSISOC, Valladolid

Dubuc S (2004) Dynamisme rural: l'effet des petites villes. L'Espace Géographique 1:69–85

Felemou M (2011) Analyse de données de flux de navetteurs. Extractions de modèles (p 31 Rapport de stage de 1ère année de Master Statistiques et Traitement de Données). Cemagref, Aubière

Fernandez LE et al (2005) Characterizing location preferences in an exurban population: implications for agent-based modeling. Plan Des 32(6):21

Fontaine C, Rounsevell M (2009) An agent-based approach to model future residential pressure on a regional landscape. Landscape Ecol 24(9):1237–54

Gargiulo F et al (2010) An iterative approach for generating statistically realistic populations of households. PLoS One 5(1):9

Gargiulo F et al (2011) Commuting network model: going to the bulk. JASSS 15(2):6

Gilbert N, Troitzsch KG (2005) Simulation for the social scientist. Open University Press, Maidenhead

Givord P (2003) Une nouvelle enquête Emploi. Economie et Statistiques 362:59–66

Goux D (2003) Une histoire de l'enquête Emploi. Economie et Statistiques 362:41–57

Grimm V et al (2006) A standard protocol for describing individual-based and agent-based models. Ecol Model 198:115–126

Grimm V et al (2010) The ODD protocol: a review and first update. Ecol Model 221:2760–2768

Holme E et al (2004) The SVERIGE spatial microsimulation model: content, validation and example applications (p 55). Spatial Modelling Centre, Sweden

Huet S et al (2011) Micro-simulation model of municipality network in the Auvergne case study. In PRIMA (ed) Working paper (p 63). IRSTEA, Aubière

INSEE division Redistribution et Politiques Sociales (1999) Le modèle de simulation dynamique DESTINIE. Série des documents de travail de la Direction des Etudes et Synthèses Economiques 124:1–49

Johansson M, Rauhut D (2007) The spatial effects of demographic trends and migration. E. P. 1.1.4., European Observation Network for Territorial Development and Cohesion (ESPON):65

Lenormand M, Huet S et al (2012a) Deriving the number of jobs in proximity services from the number of inhabitants in French rural municipalities. PLoS One 7(7):13

Lenormand M, Huet S et al (2012b) A universal model of commuting networks. PLoS One 7(10)

Moeckel R et al (2003) Microsimulation of land use. Int J Urb Sci 71(1):14–31

Morand E et al (2010) Demographic modelling: the state of the art (FP7–244557 Project Sustain City, p 39). INED: Paris

Müller K, Axhausen KW (2011) Population synthesis for microsimulation: State of the art, 90th Annual Meeting of the Transportation Research Board (Washington D.C.)

Orcutt GH (1957) A new type of socio economic system. Rev Econ Statist 58:773–97

Parker Dawn C et al (2003) Multi-agent systems for the simulation of land-use and land-cover change: a review. Ann Assoc Am Geogr 93(2):314–37

Perrier-Cornet Ph (2001) La dynamique des espaces ruraux dans la société française: un cadre d'analyse. Territoires 2020, 3 (Etudes et prospectives—Data), 61–74

Polhill JG et al (2008) Using the ODD protocol for describing three agent-based social simulation models of land-use change. JASSS 11 (2.3):30

Rindfuss R et al (2004) Developing a science of land change: Challenges and methodological issues. PNAS 101(39):13976–81

Soumagne J (2003) Les services en milieu rural, enjeu d'aménagement territorial. Revista da Faculdade de Letras—Geografia I série, XIX.

Turci L et al (2010) Provisional demographic outline (FP7–244557 Project SustainCity, p 24). INED, Paris

Ullman E, Dacey M (1960) The minimum requirement approach to the urban economic base. Pap Proc Reg Sci Ass 6(192):175–194

Verburg PH et al (2002) Modelling the spatial dynamics of regional land use: the clue-s model. Environ Manage 30(3):391–405

Verburg PH et al (2004) Land use modelling: current practice and research priorities. GeoJournal 61:309–324

Verburg PH et al (2006) Downscaling of land use change scenarios to assess the dynamics of European landscapes. Agri Eco Environ 114:39–56

Chapter 9
How to Characterise and Parameterise Agents in Electricity Market Simulation Models: The Case of Genersys

George Grozev, Melissa James, David Batten and John Page

9.1 Model Description Overview

9.1.1 Agent-Based Simulation for Electricity Markets

Agent-based modelling (ABM) is establishing itself as a mature area of study of complex adaptive systems (Samuelson and Macal 2006; Macal and North 2009). It is a methodology to explore the interplay between the micro and macro levels of dynamic systems, specifically to discover how the micro interactions between agents generate different macro behaviour and structures (Epstein and Axtell 1996; Batten 2000). Also, ABM provides a constructive framework to study spatial, temporal and network effects in systems where the corresponding individual agents are located in space, time and grids. As the ABM allows us to couple technical, social and environmental aspects of real systems into a single computational model it has been used to build several simulation models of electricity markets around the world such as EMCAS (Veselka et al. 2002; Conzelmann et al. 2010), NEMSIM (Grozev et al. 2005; Batten and Grozev 2006, 2008) and its commercial successor Genersys (Grozev et al. 2008; James et al. 2011; Sugianto et al. 2012).

G. Grozev (✉) · M. James · D. Batten · J. Page
The Commonwealth Scientific and Industrial Research Organisation (CSIRO), Melbourne, Australia
e-mail: George.Grozev@csiro.au

M. James
e-mail: Melissa.James@csiro.au

D. Batten
e-mail: David.Batten@csiro.au

J. Page
e-mail: John.Page@csiro.au

A. Smajgl, O. Barreteau (eds.), *Empirical Agent-Based Modelling – Challenges and Solutions*, DOI 10.1007/978-1-4614-6134-0_9,

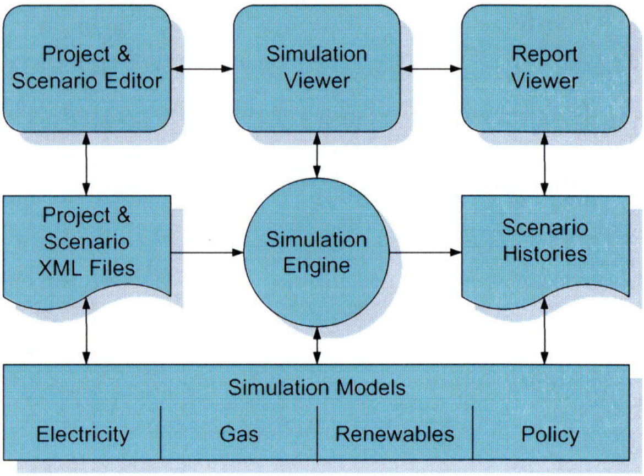

Figure 9.1 High level model structure of GENERSYS—the simulation models in the bottom row are the focus of this paper

9.1.2 Overview of Genersys

Genersys is an agent-based software application for simulating electricity and gas markets. It was developed by CSIRO's Energy Transformed Flagship in collaboration with Core Energy Group and AGL Energy using object-oriented technology and the Java programming language. It has user-friendly interfaces for defining scenarios, viewing simulations, and viewing simulation results, and it integrates multiple models which define the characteristics and behaviours of the different aspects of the energy market (see Fig. 9.1).

In the Genersys electricity model, generator companies own generating plants which are located in regions. Generator plants consist of one or more generating units, each using a particular type of technology to generate electricity, and each with its own characteristics including capacity, fuel type, efficiency, and availability. The electricity network is modelled as an inter-connected regional grid with transmission interconnectors linking the regions.

Genersys simulates the electricity market at 30 min intervals over a desired period, for example, 30 years. The regional electricity demand model within Genersys is based on a method developed by Thatcher (2007). Demand data sets are constructed at 30 minutes intervals. The data is consistent with climate change scenarios and takes into account economic growth factors. Generator companies submit bids for each generating unit to supply electricity into the market to match this demand. An independent market operator balances demand and supply, using submitted bids to determine the amount of electricity to be dispatched by each generating unit, together with the spot price for each region.

A set of bidding models are available for use in Genersys, including dynamic and adaptive bidding. Cost-based bids are frequently used for scenario development as they are based on the economic costs of supply for each generating unit. Cost-based bids factor in carbon price, fuel prices, operation and maintenance costs, generation efficiency, and minimum generation restrictions. Optionally, cost-based bidding can factor in fixed operating and maintenance costs, capital costs, and generation capacity factors.

In addition to modelling fossil-fuel based technologies, Genersys has special models for renewable generation including wind, hydro, and solar. These models are linked to climate variables such as temperature, wind speed, water inflows, and solar radiation. Mandatory renewable energy targets are also modelled.

Genersys models greenhouse gas emissions and includes a carbon pricing mechanism. A carbon price may be defined exogenously using an explicit carbon price trajectory. Alternatively an emissions target may be defined and the carbon price is endogenously determined in response to the level of emissions output by the system.

Within a Genersys simulation, as demand for electricity grows over a multi-year period, the existing generation capacity within the simulated market system may no longer be sufficient to meet this growing demand. This usually results in power outages ('blackouts'). In order to meet growing demand and avoid blackouts Genersys can simulate generation capacity expansion. This can be done either statically or dynamically.

The Genersys electricity model is integrated with the Genersys gas model. The gas model simulates gas supply and demand from gas fields, through gas pipelines and to delivery points. Delivery points include gas-based electricity generators.

To run a Genersys simulation, a scenario is defined via an easy to use graphical interface. While the simulation is running, its progress can be viewed via an interactive map and many dynamic graphs. Once the simulation has completed its run, the simulation results can be viewed via a report interface. Results include electricity production, spot prices, greenhouse gas emissions, blackouts, revenue, among others.

Alternative scenarios may be explored within Genersys. For example, two different scenarios may be defined which use different carbon price trajectories. After the two simulations have been run, results can be compared to explore the impact of carbon pricing (Fig. 9.2).

Genersys also includes a Monte Carlo framework which allows the user to run the same scenario multiple times, each time using a different random seed number. The results vary between individual runs and the mean and standard deviation results values are calculated and displayed. In this way Genersys is able to model some of the randomness inherent in the system, in particular in energy demand, outages, and wind generation.

Genersys development utilised cutting edge technologies within the open source realm to develop the application. Code quality was maintained using an automated testing and release build process. The application was unique from the technical point of view on several aspects. It was platform independent and could run on desktop as well as high performance cluster computing machines. It had a graphical user interface as well as a command-line interface.

The application was capable of managing large sized data sets. It could process and display results using its interactive scalable vector graphics and JFreeChart components[1]. All developers who worked on the project team had several years of industry experience which helped in creating a robust and scalable application.

[1] http://www.jfree.org/jfreechart/.

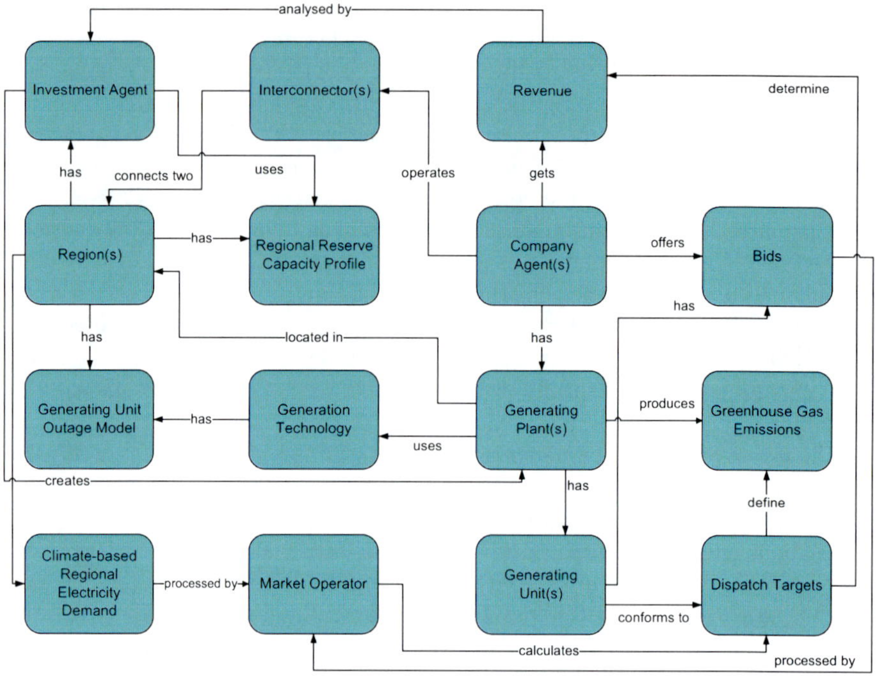

Figure 9.2 Genersys electricity model

9.1.3 *Design Concepts*

Emergence Market prices of electricity, total greenhouse gas emissions, future blackouts and future generation mix are all system-level emerging phenomena from running Genersys scenarios with different bidding strategies of generator companies, greenhouse gas policies, cost of new generation technologies or climate change impacts. The term generation mix used here describes the composition of different types of power plant technologies that supply electricity to the network. It could be expressed in terms of installed capacity (as potential) or dispatched capacity (as utilised) and it is usually measured in MW.

Adaptation Generating unit bidding can be adapted depending on external conditions (market regional price, regional demand or regional reserve capacity). Dynamic bidding is usually used in conjunction with other bidding models implemented in Genersys, for example cost-based bidding or time of the week based bidding. Another example of adaptive bidding is energy targeted bidding, which is used when a specific energy output is expected for a given period (year), which may be the case in hydro generation as the total electricity is constrained by the amount of available water. Another adaptive relationship in Genersys that can be configured is the link between carbon tax and total greenhouse gas emissions.

Fitness No explicit fitness is modelled in Genersys and no utility functions, e.g. profit maximisation for generator companies, is implemented in the simulation tool at this stage.

Prediction Complex "look-ahead" bidding strategies were experimented with at earlier development stages of the simulation tool. They required a lot of additional simulation time as each bidding company was allowed to experiment with several bidding options for a given look-ahead period of 1 week, keeping all other bids the same, before selecting the best bid in terms of profit maximisation. For one bidding company with three bidding options and 1 week look-ahead period it required 3 weeks of additional simulations and a complex roll-back mechanism.

The capacity expansion algorithm (CEA) in Genersys has an element of prediction or extrapolation of future revenue from a given class of generation technology, based on historical price distributions.

Sensing No explicit sensing mechanisms are implemented in Genersys. The carbon price setting mechanism in Genersys can respond to greenhouse gas emission levels in the system and increase/decrease the carbon price to induce behaviour that will reduce/increase greenhouse gas emissions. Other features include "trend variables" with annual modifications (e.g., growth, reduction) to consider cost and other changes over time. Both the CEA and the bidding models can utilise cost-based trend variables.

Interaction A wide range of interactions are modelled in Genersys. Firstly, it is the market interactions of generator companies through bidding offers and levels of electricity demand that define the dispatch, the market price and associated revenues. Secondly, it is the interactions between electricity supply and electricity demand that defines periods of unserved energy or blackouts. Thirdly, it is the interaction between economics and technical components of this complex market system, for example forced outages of generating units make them unavailable for generation and as a consequence not able to bring revenue to the company which owns them. At the same time an outage of a generating unit may have an impact on the regional market price. Fourthly, some elements of impacts on the environment are calculated in terms of greenhouse gas emissions released due to electricity generation and amount of different fuels used.

Stochasticity Some models in Genersys have stochastic variables. For example the timing and duration of forced outages of generating units are defined based on random number distributions for mean time to fail and mean time to repair. To account for stochasticity introduced by a number of models, Genersys has a Monte Carlo simulation framework that allows multiple simulations to run with different random number seeds and to estimate the variability of the simulation outputs.

Collectives There is an ownership structure of generating plants across market regions and grouping of generating plants in regions based on geographical locations and definitions of market regions. Several types of market participants (e.g., generators, network transmission companies, etc.) are modelled differently, however, no grouping within each type is modelled.

Observation The Australian Energy Market Operator[2] (AEMO) (previously NEMMCO) reveals a significant amount of market data on 5 minutes, 30 minutes, daily, weekly, monthly and annual basis about the NEM. Time series market data at 30 minutes intervals from the NEM is utilised by the simulation tool and represented as historical data, for example market prices and demands, bids, interconnector capacities, etc. Some of these time series may be configured to be used as inputs if a simulation is run for past periods, helping with the model assessment.

9.2 Overview: Framework Specific Sequence

Here we summarise the development process of Genersys using the decision tree and the characterisation and parameterisation (CAP) framework described in Chap. 1 and also based on the parameterisation methods described by Smajgl et al. (2011).

Because Genersys involves relatively small populations, nearly 100% of the population that needs to be simulated can be accessed and modelled based on market and other types of data. This makes the up-scaling step obsolete and Genersys matches reasonably well Case 12 from the classification given in Chap. 1.

The NEM spans five states and one territory, namely New South Wales, Victoria, Queensland, South Australia, Tasmania and the Australian Capital Territory (ACT) (AEMO 2010). Genersys models all five states as regions (the ACT is included in NSW). One hundred and sixty five market participants were registered in the market in 2012, including 83 generator companies, 9 transmission network operators, distribution companies and traders. Only scheduled generator companies, wind generator companies and transmission companies are modeled in Genersys, plus the market operator. One investment company could be configured for each market region. There are about 355 generating units in the market (the physical units are many more as in some cases several small units are aggregated in one logical unit for market purposes). Only big, scheduled and wind generating units are considered by Genersys.

9.2.1 M1 Model Characterisation Methods

Agent classes and principle behaviours were identified based on literature review, expert knowledge, participant observation, interview and meetings with electricity industry experts, a focus group meeting and through analysing market data. Ongoing relationships with industry-based development partners in the latest develop-

[2] www.aemo.com.au.

ment stage helped the research and development team to advance the tool quickly and to address complex user needs.

9.2.2 M2 Attribute Data Elicitation Method

Agent attributes were elicited and initialised based on analysis of many market, research and third party data sets.

9.2.3 M3 Behavioural Data Elicitation Method

Agent behavioural response data were obtained through interviews, utilising market and third party data and implementing a Monte Carlo simulation framework to account for stochasticity.

These three steps (M1, M2 and M3) and the model assessment step, informed each other and were performed iteratively, building several versions of the simulator, with much revising, differentiating and expanding attribute data or behavioural response data throughout the process.

9.2.4 Model Assessment

Model assessment included comprehensive testing at different levels and validation of distributions of simulation outputs against real market data. Many improvements were implemented based on feedback from industry experts—users of the tool.

9.3 Technical Details

Genersys was developed using an agile software development methodology based on iterative, incremental and adaptive development. Development steps included many builds and several major releases: (1) initial prototype of electricity model; (2) enhancement to include climate change consistent electricity demand model; (3) mature electricity model, including advanced graphical user interface based on scalable vector graphics functionality; (4) first commercial version of the model, including a gas model; (5) second commercial version, including renewable generation (wind and solar), capacity expansion algorithm and policy options.

At each of these stages a variety of methods were used to formulate the model (or extension to the model), identify agent classes and the structure of agent behaviours, and to model the environment in which they operate. These methods are described below.

9.3.1 M1 Methods

Methods M1 are usually associated with the identification process of distinct agent classes, model structures and sequences of action (Smajgl et al. 2011).

9.3.1.1 Literature Review

As in many other research situations, a literature review of electricity market modelling issues and research questions helped to identify some important contributions and trends in this area at the beginning of the current century (see contributions from Vlahos et al. 1998; Veselka et al. 2002; Bower and Bunn 2001). This period coincided with rapid changes and restructuring of the electricity industry in many countries around the world aiming to introduce higher efficiency and competition in this sector. Traditionally the electricity sector was vertically integrated and government owned and it was rapidly privatised and disaggregated horizontally (by service area) and vertically (by businesses of generation, transmission, distribution and retail). Many new company players appeared at the market place. The traditional modelling and simulation approaches used in the last several decades in the previous century were not adequate to capture this new complex reality of competing, adaptive and profit-maximising market players. Our review had identified the Electricity Market Complex Adaptive System (EMCAS) developed by Argonne National Laboratory in U.S.A. (Veselka et al. 2002) as a potential agent-based tool to be adapted for the modelling purposes of Australia's NEM, however, there were organisational, ownership and financial difficulties in adoption of this tool. In this context CSIRO Energy Transformed Flagship decided to develop its own tool in an incremental way. At each major development step scientific literature was explored to identify previous solutions, models and in many cases—to identify data availability.

9.3.1.2 Meetings with Energy Industry Experts

A number of meetings with electricity industry experts were initially organised by the modellers to get a better understanding of the simulation needs of the industry and to identify which business areas required a major modelling focus. Some of these meetings took place a few years after the introduction of Australia's National Electricity Market in December 1998. This was a major restructuring of the previous centralised delivery of electricity to customers by vertically integrated utilities; the previously used simulation and optimisation models were not sufficient anymore and the situation created an opportunity for developing an agent-based market simulator.

9.3.1.3 Focus Group Meeting

A focus group meeting between researchers and experts from the electricity industry and government was held at the initial stage of the project. A number of industry and government experts were invited to participate by sending them invitation letters and a specially prepared brochure about the simulation tool. At the same time notification letters were sent to about two dozen company leaders—chief executive officers, general managers and executive directors of energy companies (generators, retailers, network transmission and distribution companies), the market operator, consultancy companies, government departments and relevant associations. The company leaders were invited to nominate participants from their organisations. The total number of participants, including modellers, was 40. The focus group meeting is a formal method of capturing the knowledge of the experts, in this case in the area of electricity markets. It was a half-day event with intense interactions between the experts and modellers. Initially the concepts and key features of the proposed simulation tool were explained by the research team. A demonstration of the initial simulation prototype was presented. The experts had group discussion and provided feedback to modellers about gaps in functionality of existing models and tools for electricity markets adopted at the time. The focus group meeting was able to identify strategic directions, key functionalities and modelling issues of high importance for the industry and stakeholders.

9.3.1.4 Analysis of Market Data

Market data from AEMO was analysed (Hu et al. 2005) aiming to help with identification and selection of classes and model structures. As an Independent System Operator of the national electricity and gas markets, AEMO generates big data sets on a daily basis. This information reveals many parameters and behaviours of generator companies for example. It required a well defined selectivity in terms of modelling classes and their attributes as the complexity of the real markets is huge and impossible to recreate in a reasonable simulation system. A decision was taken to model the wholesale electricity market, but not several other markets that are related to power system security and reliability services. AEMO operates eight separate markets (apart from the wholesale electricity market) for delivery of frequency control, network control and network restart ancillary services. These eight markets are very important from a technical point of view, however, the traded quantities and corresponding dollar values are insignificant in terms of long-term simulation models and they were not included in the Genersys simulation model.

9.3.1.5 On-going Relationship with Users and Potential Users of the Simulation Tool

When Genersys was commercialised in 2006 with funding from CSIRO Energy Transformed Flagship, Federal Government, Core Energy Group and AGL Energy, close relationships between the three development organisations were created. The new users from Core Energy and AGL Energy provided valuable feedback and detailed specifications for new entities and functions that elevated the Genersys integrated modelling capability to a new level. Several features and improvements were implemented. AEMO invited the development team to participate in tenders for simulation capabilities and through this process informed new desired features and capabilities. Several other organisations received trial versions of Genersys and provided valuable feedback to the research and development team.

9.3.2 M2 Methods

Methods M2 are usually associated with specification of values for agent attributes and various data sources for initialisation of these attributes.

9.3.2.1 Use of Market Data

Genersys can be populated with data from a variety of sources, however, market data from the independent system operator AEMO is the main input stream for the simulator. As indicated in the first Section, AEMO publishes a significant amount of market data on 5-min, 30-min, daily, weekly, monthly and annual basis about the NEM performance and predictions. Market historical data about electricity prices, demand, generator bids and capacity flows are packaged in Genersys and can be presented in the simulation screens if a simulation is performed in the past. In some cases market historical data could be used as input for simulations as well. There are several file options that allow historical data to be used for future projections, for example future electricity demand.

AEMO and other agencies produce a variety of technical reports (for example, AEMO's annual National Transmission Network Development Plan), which contain data about each region, interconnector and generating unit including capacity, technology, costs, fuel type, efficiency, availability, greenhouse gas emissions, and outages. Reports also include planned generation expansion data. This data is used to assign values to the attributes of regions, interconnectors, generating units, generating plants, generation technologies, generation expansion agents, etc. AEMO's registration file is used to initialise company agents and also has the latest information about all generating plants and generating units in the market. AEMO's 30 minutes demand data series are used by the Genersys electricity demand model (Thatcher 2007) to configure daily demand patterns and project future demand values. Economic growth rates are based on AEMO's 2010 Electricity Statement of Opportunities.

AEMO bidding data was analysed, processed, and used to populate the bidding models and for "backcasting" of simulations. Generation capacity is based on AEMO's *Electricity Statement of Opportunities* (AEMO 2011).

The simulation includes a mandatory renewable energy target (MRET). The values used were obtained from the Clean Energy Regulator website[3].

9.3.2.2 Use of Research Data

In the simulation, electricity demand is calculated based on a linear regression model described by Thatcher (2007). It uses climate data (air temperature and relative humidity) derived from the CSIRO Mk3.0 coupled atmosphere-ocean General Circulation Model (GCM) after being dynamically downscaled to 60 km resolution by the CSIRO Conformal Cubic Atmospheric Model (CCAM) (McGregor 2004) as part of a future climate projection under the Intergovernmental Panel on Climate Change (IPCC) Special Report on Emissions Scenarios (SRES) A2 scenario (IPCC 2000).

Climate data (wind speed) is used also for the wind generation model.

The CEA uses projections of the future costs of electricity generation technologies, which are generated by CSIRO's Global and Local Learning Model (GALLM) (Hayward et al. 2011).

To make use of research data from other tools and models, customised interfaces to some of them can be designed and implemented. Genersys has a specialised interface to feed via a specially formatted file, output data from another research model—the Energy Sector Model (ESM) (CSIRO 2009), which CSIRO co-developed with the Australian Bureau of Agricultural and Resource Economics (ABARE). The output from ESM can be converted to generating plant configurations for new generating units to be commissioned into the future.

9.3.2.3 Use of Other Types of Data

A variety of additional economic, environmental (greenhouse gas emission intensity factors of generation technologies and generating plants) and other types of data sourced from technical reports and web sites are utilised by the simulation tool.

It must be noted that although Genersys can be populated with data from the sources described above, it is up to the user of Genersys to decide exactly how they wish to populate the model prior to running the simulation. In using Genersys for 'what-if' scenario-based simulations, the user can experiment with using a variety of data input sources.

[3] http://ret.cleanenergyregulator.gov.au/.

9.3.3 M3 Methods

Methods M3 are usually associated with parameters of behavioural rules or responses that agents follow. As Genersys has many modules implementing different models, here we briefly discuss only two groups of them related to stochastic parameterisation and system evolution. Each of these groups is illustrated with examples.

9.3.4 Stochastic Parameterisation

9.3.4.1 Monte Carlo Framework

In order to model stochastic processes within Genersys a Monte Carlo framework was developed based on introducing random number parameters within the simulation. These random parameters were injected within the computation algorithms of some of the core models. The core models and the action of introducing the random parameter within them are listed in the Table 9.1 below.

By starting each simulation run with a different random number seed, each simulation run basically simulates a different set of market conditions. The different conditions create a variance in demand, dispatch and spot prices. The Genersys Monte Carlo framework was extended to run on an IBM ‘Blade’ computer cluster. The cluster based implementation was able to perform the simulations in parallel and compute the mean and standard deviation of all key attributes for the reporting process in acceptable time periods in the case of 20-year scenarios with 50 different simulation runs. Thus, the Monte Carlo implementation on the IBM cluster server offered a significant boost to the analytical capabilities of Genersys.

9.3.4.2 Wind Generation

Electricity output from renewable generation such as wind generation is usually difficult to forecast due to complex wind speed patterns. It is even more complex to model electricity output from several wind farms located in different locations of the same geographical region as the wind speed in these locations may be partially correlated due to predominant weather patterns. To address some of these issues, the research and development team of Genersys has constructed a parameterisation method for simulating wind farms with wind speeds that have realistic probability distributions, intra-day variability and correlation with other wind farms (Thatcher 2009). Once the wind speed time series are constructed for each location they are then converted into electricity generated by the wind farm so that the effect of wind generation on the market can be realistically investigated. A feature of the wind farm

Table 9.1 Random parameter use in Genersys

Model	Action of random parameter
Generating unit	To set the start and end of the outages and to decide whether the next outage will be forced or planned
Interconnector	To set the start and end of the outages
Seasonal demand calculator	Computes the demand based on seasonal variance
Wind model	Calculate wind speed
Simple bid generator	To generate the price matrix for the bids

parameterisation is that it is designed to be compatible with the climate datasets used for electricity and gas demands, as derived from the CCAM (McGregor 2004). As a result, the wind model is able to simulate the electrical output from wind farms over a long period (40–50 years) in 30 minutes intervals.

9.3.4.3 Outage Models

Sometimes availability of data may drive a specific model implementation. A good example is the forced outage model for generating units. In order to keep company data confidential, AEMO aggregates and only then publishes outage data for generating units in each market region based on three types of generation: baseload, intermediate and peaking. Parameters such as mean time to fail and mean time to repair are provided per region and for these three types of generation. In order to be properly initialised, the Genersys forced outage model was designed to account for these three types of generation in each region. In contrast, as outage data for transmission lines in Australia is not rich, data sets from U.S.A. were used.

9.3.5 System Evolution

For a long-term simulation it is important to consider growth in the number of some and reduction of other simulation entities. For example both electricity supply and demand may grow into the future due to population and economic growth. Due to development of disruptive new technologies, it is likely also that future electricity demand may stabilise or significantly reduce. Cost and other parameters also change with time and they don't necessarily grow linearly or by equal annual steps. In Genersys there are a number of ways to contribute to the system evolution and we briefly explain the capacity expansion algorithm (CEA), the electricity demand model and trend variables.

9.3.5.1 Capacity Expansion Algorithm (CEA)

To increase future electricity supply and to modify the existing generation mix a functionality to create new generating plants into the future is necessary. In Genersys this can be done in two ways—statically and dynamically. Using the static method the user can configure a generating plant to begin generating in a specific moment in the future simulation time. In this case it is assumed that the user has all the necessary information about this new generating plant in terms of generation technology, size, time and location of investment, etc. In the dynamic case of the CEA, which is implemented in Genersys, several generation technologies per a given region are evaluated for investment in new generating plants as a response to growing electricity demand, unserved energy and/or high wholesale prices of electricity. The CEA considers capacity factors associated with peak, intermediate and base load generation, technology and fuel costs, carbon price, and distribution of recent electricity prices. It is a backward looking algorithm that does not require complex "look-aheads" to evaluate future cost and revenues subject to behaviour of other agents. However, as a fully integrated simulation algorithm within Genersys, it can be used to assess complex investment scenarios. The CEA creates modular expansions—a single generating unit with predefined generation capacity of the best ranked generation technology in a region will be invested in if the investment criteria are satisfied. Generating plants newly created by the CEA begin to participate in the market, producing and selling electricity. They will have immediate impact on the balance between demand and supply and ultimately will influence the electricity market prices as well. The revenue received by a new generating plant allows evaluation of its financial viability over the lifetime of the plant or for a specified simulation period. The CEA can be used to evaluate investment decisions in a mix of generation technologies, including renewable technologies such as wind and big solar generation.

9.3.5.2 Electricity Demand

Any regional electricity demand in the NEM has highly complicated daily, weekly and seasonal patterns as an aggregation of electrical uses by thousands and millions of residential, commercial and industrial users at any given time period. In order to be able to simulate future electricity demand for any region in the market, Genersys is equipped with a specialised method for constructing probable daily electricity demand datasets (in 30 minutes intervals) that are consistent with climate change predictions (Thatcher 2007). The model correctly reproduces the complex demand patterns and it is based on Cooling Degree Days (CDD) and Heating Degree Days (HDD). It is able to predict detailed changes to load duration curves as a consequence of a 1 °C increase in average temperatures of Australian capital cities.

9.3.5.3 Trend Variables

A trend variable can define variations of a given parameter or attribute over time. Trend variables are useful in modelling any parameter that changes such as carbon price, emission target, reservoir capacity or economic growth. It allows users to set changes at specific future dates or annual increments.

9.4 Lessons/Experiences

This chapter describes the process of characterisation and parameterisation of computer agents in Genersys—a simulation tool for electricity markets with focus on Australia's National Electricity Market (NEM). A brief introduction of Genersys is provided in the first section, including some agent-based design concepts, e.g., emergence, adaptation, interaction, etc. Because Genersys involves a relatively small number of company agents, nearly 100 % of the agent population that needs to be simulated can be accessed and modelled based on market and other types of data. This makes the up-scaling step (usually required in many other models with a significant number of agents) obsolete and Genersys matches reasonably well Case 12 from the classification given in Chap. 1. Three classes of characterisation and parameterisations are described in the technical section—model, attribute and behaviour data elicitation methods. Through initiatives such as formal focus group meetings, gathering observations of industry experts, analysing market data and a selective approach in representing real systems, modellers can improve the design and potential future use of their simulation systems.

9.4.1 Focus Group Meeting and Gathering Observation of Experts

The focus group meeting between researchers, modellers and experts from the electricity industry and government was very useful for strategically aligning the planned simulation capability of Genersys to industry and practitioners' needs, specifically to address longer-term market developments and evolution. Meetings between the research team and industry and government experts were critical to get proper understanding of a variety of system issues to be modelled from perspectives of different businesses and organisations.

9.4.2 Predictability, Accurate Representation and Usability

Axelrod (1997) explains the diverse purpose that social simulation can play through prediction, performance, training, entertainment, education, proof and discovery.

The author makes the point that simulation modelling can be used to add our intuition and that the purpose of the agent-based modelling is to enrich our understanding of fundamental processes not necessarily to provide an accurate representation of a given real system.

In the case of Genersys, the roles of prediction, proof and discovery are much more essential than the others in the list. Through the development process of this simulation tool our focus has shifted more towards the prediction aspects and accurate representation in contrast to simplicity of agent representation and discovery of fundamental properties of the modelled system. This tendency was driven partially by our external development partners who sought practical business applications by a validated simulation model and partially by our integrated approach aiming to account for multiple interactions and multiple component models in this complex socio-economic system. As a result Genersys grew to a complex simulation model with multiple inputs, including a variety of data files and is able to report on approximately 800 model attributes. The model requires careful preparation of inputs of any simulation scenario and comprehensive training for the user. In some cases a dedicated staff resource was necessary to develop and keep the simulation knowledge in order to run and utilise Genersys. At the same time it was difficult to close the gap between the real market and the simulated model that aims to represent it, because the overall market behaviour is constantly changing, but it is very difficult to imbue the agents with the appropriate learning or adaptation characteristics to generate such macroscopic changes.

9.4.3 Use of Modern Software and Open Source Libraries

Genersys is built using modern software development technologies such as Java programming language, object oriented design and using comprehensive libraries for processing, visualisation and graphing. It has an attractive graphical user interface with an interactive map, many dynamic graphs and rich graphical reports. The Genersys software links to several free and open source Java libraries. Examples include Apache Batik SVG Toolkit[4], JFreeChart by Object Refinery and Swinburne Simulation ToolKit (SSTk) developed by Swinburne University of Technology.

9.4.4 Interpretation and Validation of Simulation Results

Genersys models decision making and adaptation of profit-driven companies in a competitive electricity market. The real market itself reveals a lot of information about market participants, their bidding, dispatch and potential revenue given the electricity prices. However, interpretation of simulation results in the case of Genersys had its own additional challenge in relation to the need to hide the identity of any single agent or company that has been characterized in a pre-determined way—

[4] http://xmlgraphics.apache.org/batik/.

e.g. through use of their historical bidding strategies or their revealed strategies during focus group or other meetings with experts.

The model was validated by comparing simulation results against historical data, and if necessary, adjustments in input parameters were made to the model. The model was also tuned and enhanced by many checks carried out by the users of the licensed organisations.

Many authors would argue that one of the main purposes of ABM is to aid our intuition (Axelrod 1997) not accurate representation of real systems. Thus validation is a particularly vexing issue. Because the learning of individual agents and the collective market behaviour is constantly changing, traditional validation methods are less appropriate for ABM. There is some value in "backcasting", but only if we know that the market is in a relatively static or predictable phase. Realistic representation of many details to pinpoint accuracy is unnecessary, and may even be counter-productive. Given the open-endedness of ABM, what is more important is that validation methods focus on how closely the simulated macro behaviours or outcomes reflect the observed ones in a qualitative sense. However, the industry users had a strong expectation of predictability and accuracy and tended to insist on very complicated validation exercises. For the industry users the paradigm of ABM was considered as an add-on for representing decision making in a simulation system that has a very high level of accuracy and predictability, an expectation, which was very challenging to realise.

Acknowledgements Genersys was developed by CSIRO Energy Transformed Flagship in collaboration with Core Energy Group and AGL Energy. The authors would particularly like to thank Geoff Lewis, Tennyson Wickham previously of Core Energy Group and Per da Silva—previously of CSIRO for their valuable contribution to the development of Genersys.

References

AEMO (2010) An introduction to Australia's National Electricity Market. http://www.aemo.com.au/en/About-AEMO/~/media/Files/Other/corporate/0000-0262%20pdf.ashx. Accessed 24 July 2013

AEMO (2011) Electricity Statement of Opportunities. http://www.aemo.com.au/Electricity/Planning/Electricity-Statement-of-Opportunities. Accessed 24 July 2013

Axelrod R (1997) Advancing the art of simulation in the social sciences. In: Conte R, Hegselmann R, Pietro Terna P (ed) Simulating social phenomena p. 21–40. Springer-Verlag, Berlin

Batten DF (2000) Discovering artificial economics: how agents learn and economies evolve. Westview, Boulder

Batten D, Grozev G (2006) NEMSIM: finding ways to reduce greenhouse gas emissions using multi-agent electricity modelling. In: Pascal P, Batten D (ed) Complex science for a complex world: exploring human ecosystems with agents. ANU E, Canberra

Batten DF, Grozev GV (2008) Managing energy futures and greenhouse gas emissions with the help of agent-based simulation. In: Ruth M, Davidsdottir B (ed) Changing stocks, flows and behaviors in industrial ecosystems. Elgar, Cheltenham.

Bower J, Bunn D (2001) Experimental analysis of the efficiency of uniform-price versus discriminatory auctions in the England and Wales electricity market. J Econ Dyn Control 25(3–4):561–592

Conzelmann G, Boyd G, Koritarov V, Veselka T (2005) Multi-agent power market simulation using EMCAS. Power Engineering Society General Meeting, 12-16 June 2005, 3:2829-2834. doi:10.1109/PES.2005.1489271

CSIRO (2009) Energy sector modelling. http://www.csiro.au/en/Organisation-Structure/Flagships/Energy-Flagship/Energy-Sector-Model.aspx. Accessed 24 July 2013

Epstein JM, Axtell R (1996) Growing artificial societies—social science from the bottom up. MIT Press, Cambridge

Grozev G, Batten D, Anderson M, Lewis G, Mo J, Katzfey J (2005) NEMSIM: agent-based simulator for Australia's National Electricity Market. SimTecT 2005 Conference Proceedings, Sydney, Australia

Grozev G, James M, Thatcher M, Lewis G, Wang C, da Silva P (2008) GENERSYS 2.0 User Guide. CSIRO Technical Report, 203 pages

Hayward JA, Graham PW, Campbell PK (2011) Projections of the future costs of electricity generation technologies: an application of CSIRO's Global and Local Learning Model (GALLM). CSIRO technical report.

Hu X, Grozev G, Batten D (2005) Empirical observations of bidding patterns in Australia's National Electricity Market. Energ Policy 33(16):2075–2086

Intergovernmental Panel on Climate Change (IPCC) (2000) IPCC Special Report Emissions Scenarios

James M, Grozev G, Page J, Yum K-K (2011) Using Genersys to model electricity generation expansion. 19th International Congress on modelling and simulation MODSIM 2011, Perth, Australia, 12–16 December 2011. http://www.mssanz.org.au/modsim2011/C4/james.pdf. Accessed 14 March 2012

Macal CM, North MJ (2009) Agent-based modelling and simulation. In: Rossetti MD, Hill RR, Johansson B, Dunkin A, Ingalls RG (eds) Proceedings of the 2009 Winter Simulation Conference, Austin, Texas, 2009

McGregor JL (2004) C-CAM: geometric aspects and dynamical formulation. CSIRO Marine and Atmospheric Research technical paper No. 70

Samuelson D, Macal C (2006) Agent-based simulation comes of age. OR/MS Today, 33(4): 34–38, August 2006. Lionheart Publishing, Marietta, GA, USA. http://www.orms-today.org/orms-8-06/fragent.html. Accessed 14 March 2012

Smajgl A, Brown DG, Valbuena D, Huigen MGA (2011) Empirical characterisation of agent behaviours in socio-ecological systems. Environ Modell Softw 26(7):837–844

Sugianto LF, Grozev G, Liao Z, James M, Page J (2012) Simulating cost-based bidding models in Genersys. 6th Int. Conference on Soft Computing and Intelligent Systems, 20–24 November 2012, Kobe, Japan.

Thatcher MJ (2007) Modelling changes to electricity demand load duration curves as a consequence of predicted climate change for Australia. Energy 32(9):1647–1659. doi:10.1016/j.energy.2006.12.005

Thatcher MJ (2009) A stochastic wind farm parameterisation for Genersys. CSIRO technical report.

Veselka T, Boyd G, Conzelmann G, Koritarov V, Macal C, North M, Schoepfle B, Thimmapuram P (2002) Simulating the behavior of electricity markets with an agent-based methodology: the electricity market complex adaptive system (EMCAS) model, International Association for Energy Economists (IAEE) North American Conference, 6–8 October, 2002, Vancouver, BC

Vlahos K, Ninios P, Bunn D (1998) An integrative modelling approach for understanding competitive electricity markets. J Oper Res Soc 49(3):187–199

Chapter 10
An Agent-Based Model Based on Field Experiments

Marco A. Janssen

10.1 Introduction

This chapter described the empirical calibration of a theoretical model based on data from field experiments. Field experiments on irrigation dilemmas were performed to understand how resource users overcome asymmetric collective action problems (Janssen et al. 2012). The fundamental problem facing irrigation systems is how to solve two related collective action problems: (1) the provision of the physical and ecological infrastructure necessary to utilize the resource (water), and (2) the irrigation dilemma where the relative positions of "head-enders" and "tail-enders" generate a sequential access to the resource itself (water). If actors act as rational, self-interested, agents, it is difficult to understand how irrigation infrastructure would ever be constructed and maintained by the farmers obtaining water from a system as contrasted to a government irrigation bureaucracy. Wittfogel (1957) argued that a central control was indispensable for the functioning of larger irrigation systems and hypothesized that some state-level societies have emerged as a necessary side-effect of solving problems associated with the use of large-scale irrigation.

Even if the initial problem of providing the infrastructure were solved, water that is available to the head-enders may not necessarily be shared with the tail-enders, as long as the head-enders have a positive marginal return on the use of water. The vulnerability of irrigation system performance to the behavior of self-interested rational actors leads to the question of why so many self-organized irrigation systems exist and persist for so long (Hunt 1988; Lansing 1991; Ostrom 1992).

The field experiments held in Colombia and Thailand show that trust and fairness are key components that drive the decision making. The more groups experience inequality in contributions and collection of the common resource in previous rounds, the more likely that they lower their contributions. The fact that some groups were irrigators or students had no significant effect on the decisions. We

M. A. Janssen (✉)
School of Human Evolution and Social Change, Arizona State University, Tempe, AZ, USA
e-mail: Marco.Janssen@asu.edu

A. Smajgl, O. Barreteau (eds.), *Empirical Agent-Based Modelling – Challenges and Solutions*, DOI 10.1007/978-1-4614-6134-0_10,

Table 10.1 Water production as a function of units invested in the public infrastructure

Total units invested by all 5 players	Water available
0–10	0
11–15	5
16–20	20
21–25	40
26–30	60
31–35	75
36–40	85
41–45	95
46–50	100

use the experimental data to develop a decision theoretic model of the irrigation dilemma. The model is calibrated on the data from 32 experiments. The resulting model can be used to define new hypotheses and design new field experiments.

10.2 Experimental Design

The field experiments were designed to be implemented in the field with participants who manage natural resources in their daily lives (see also Janssen et al. 2012). In the irrigation game participants have positions A, B, C, D or E. A has the first choice to harvest water from the common resource. Then B has the next turn to harvest water from whatever amount was left by A, and so on. The location of the five players is randomly determined before the first round and remains fixed over the first set of ten rounds of the game. Participants receive an endowment ω of 10 tokens in each round. First each participant makes a decision x_i on how much to invest in a public fund that generates the infrastructure and therefore determines the amount of water available for the whole group to share. In Table 10.1, the water provision generated is defined as a function f() of the total investments of the five participants.

Second, each player, in sequential turns from upstream to downstream players decides how much to extract from the available water to her, that is, the water produced minus the water extracted by those before in the sequence. Each token kept (not invested) in the first stage has a monetary value for the player that is equal to the value of each unit of water extracted in the second stage.

This experiment includes a first dilemma of upstream participants who need the contribution of downstream participants to maintain the structure of their common resource, which is crucial for the production of water in the game. However, the downstream participants can only obtain benefits from the resource if upstream participants avoid the temptation to deplete the common resource and leave little water for downstream players.

Under this asymmetric game, participants first experience a provision dilemma in the contributions stage, and then face a resource appropriation dilemma when they extract from the generated resource. The earnings of the participants are the result of provision—x_i—and extraction—y_i—decisions, and the resulting payoff z_i for player i is defined as

$$z_i = \omega - x_i + y_i, \tag{10.1}$$

where

$$\sum_{i=j+1}^{5} y_i \leq f\left(\sum_{i=1}^{5} x_i\right) - \sum_{i=1}^{j} y_i \tag{10.2}$$

for $j = 0, 1, 2, 3$, and 4.

If participants were rational self-interested individuals, nobody would invest in providing the infrastructure in the first round. Since the upstream participant is expected to collect the whole resource, downstream participants will not invest. For participant A there is no benefit to invest when others don't. If this is the reasoning of the participants in the last round of experiment we find via backward induction that the same happens for all earlier rounds. Thus, the Nash equilibrium for this game is that no one invests and all receive 10 tokens for group earnings of 50 tokens.

To define the cooperative solution, we calculate the maximum amount of the infrastructure plus tokens not invested. There are multiple social optimum outcomes. For a 41 tokens investment, a resource of 95 tokens is generated, and for a 46 tokens investment a resource of 100 tokens is generated in each round. The total earnings of the group in the cooperative solution amounts to 104 tokens, doubling the social earnings of the Nash equilibrium.

10.3 Model Description

10.3.1 Purpose

The purpose of the model is to understand which components of a decision theoretical model are most important to explain data from field experiments.

10.3.2 State Variables and Scales

The model has five agents who make decisions in 10 rounds of an irrigation experiment.

Table 10.2 Range of parameters and mean and standard deviation of distributions of parameter ranges

Parameter	Description	Range	Range (mean)	Range (standard deviation)
α	Strength aversion to exploiting others	$[\beta, 1]$	$[-1, 1]$	$[0, 1]$
β	Degree of altruistic tendency	$[-\infty, \alpha]$	$[-1, 1]$	$[0, 1]$
λ	Parameter to define probabilities	$[0, \infty]$	$[0, 5]$	$[0, 1]$
η	Initial level of cooperation of others	$[0, 1]$	$[0, 1]$	$[0, 1]$
τ_1	Learning rate investments	$[0, 1]$	$[0, 1]$	$[0, 1]$
τ_2	Learning rate extractions	$[0, 1]$	$[0, 1]$	$[0, 1]$

10.3.3 Process Overview and Scheduling

Each round all agents make first a decision how much to invest in the public fund based on expectations and preferences they have. When the level of the resource of the irrigation system is known, agents make decisions how much to take starting from the upstream agent A to the downstream agent E. The outcome of the decisions affects the expectations that are used in the following round.

10.3.4 Design Concepts

- *Prediction.* Agents have expectation on the level of cooperation of others. They have an initial level of expectation which is updated each round.
- *Interaction.* Agents interact indirectly via decisions on how much to invest in a public fund and how much to extract from a common resource.
- *Stochasticity.* Decisions are made based on the expected utility of the different options. The option that leads to the highest utility is not automatically chosen since agents make a probabilistic choice using the expected utility of the different options.

10.3.5 Initialization

The decision theoretical model that is used to define the decisions includes six parameters. Each agent has parameter values that are drawn from Gaussian distributions which parameter values are listed in Table 10.2.

10.3.6 Input

The model uses two sets of input data. Table 10.1 defines the production function of the water production. Depending on the investment decisions agents make, a certain level of the common resource—water—is produced.

The second set of data as listed in Table 10.2 is the set of parameter values of the distributions that define the decision making of the agents. The values of the parameters are determined in the calibration process as described below.

10.3.7 Submodels

10.3.7.1 Utility

We assume that agents maximize their utility. This utility u_i is formalized in a general way to include different types of other regarding preferences:

$$u_i = z_i - \alpha_i \cdot \max(z_i - \bar{z}_{-i}, 0) + \beta_i \cdot \max(\bar{z}_{-i} - z_i, 0) \tag{10.3}$$

where α and β are initially assumed to be the same for all agents, z_i is agent i's earnings, and $\bar{z}_{-i}$ is the average earnings of the other agents in the group. α can be regarded as the strength of an individual's aversion to exploiting others, and β can be regarded as an individual's degree of altruistic tendency. A lower value of β compared to α implies that a player gives a larger weight to his own payoff when his payoff is smaller than the average payoff of others compared to when it is larger. In line with Charness and Rabin (2002), we can define the following cases for $\beta \leq \alpha \leq 1$:

Case 1: The players like to have their payoffs higher than those of the other players. When $\beta \leq \alpha \leq 0$, the player is highly competitive.

Case 2: Players prefer the payoffs among all players to be equal. This "Inequity Aversion" holds when $\beta < 0 < \alpha \leq 1$ (see Fehr and Schmidt 1999).

Case 3: The third model is the so-called "Social Welfare Consideration," which holds when $0 < \beta \leq \alpha \leq 1$. The parameter a captures the extent to which a player weighs the average payoffs of the other $n-1$ agents compared to his own payoff, when his own payoff is higher than the average payoff of the others.

Case 4: If $\alpha = \beta = 0$, we have the condition that a player cares only about his or her own welfare.

10.3.7.2 Investment Decision

An agent makes two decisions. First, all agents independently make a decision how much to invest x_i. In order to make this decision, agents are assumed to estimate the expected utility based on expected behavior of others.

The expected investment level of others is equal to

$$\hat{x}_{-i} = \omega \cdot 4 \cdot \eta_i \tag{10.4}$$

where η_i is the cooperation level of other agents as expected by agent *i*. This enables agents to estimate each an expected level of the public infrastructure, $\hat{p}_i$. For each level of investment x_i, the expected level is

$$\hat{p}_i = x_i + \hat{x}_{-i} \tag{10.5}$$

Agents make a prediction how much of the resource would be available to the group using the production function of Table 10.1 with the expected value $\hat{p}_i$.

How much is expected to be available to agent *i* depends how much upstream agents have taken from $\hat{p}_i$. The lower the level of cooperation they expect from the other participants, representing here the upstream participants, the less she expects to receive from the resource before it is her turn. Hence agents assume that an amount $\hat{y}_i^A$ is available for agent *i*.

$$\hat{y}_i^A = \hat{p}_i \cdot \left(1 - \left(\frac{i-1}{5}\right)^{(2-\eta_i)}\right) \tag{10.6}$$

If agent *i* expect other agents are cooperative, $\eta_i = 1$, they take an equal share from the resource. If they are expected to be less cooperative, more than an equal share is expected to be taken.

In rounds 2 to 10 a simpler estimation technique is used by the agent to determine $\hat{y}_i^A$. The agents are assumed to expect the upstream participants take a share s_i from the expected resource size.

$$\hat{y}_i^A = \hat{y}_i^A \cdot s_i \tag{10.7}$$

The value of s_i is updated each round as defined below.

We use the values of α_i and β_i to define how much the agent takes from the share that is expected to be available to her. Agents who are selfish are expect to take the whole amount of available resources, but those with other regarding preferences are expected to take a lower level.

$$\hat{y}_i = \hat{y}_i^A \cdot (1 - \alpha_i) \cdot (1 - \beta_i) \tag{10.8}$$

Now the agent can define her utility of investing x_i and receiving $\hat{y}_i$ from the resource. Using the expected earnings, we can estimate the expected utility for agent

i for each level of investment. Based on the expected utility levels, agents make a probabilistic choice how much to invest

$$\Pr(x) = \frac{\exp(\lambda \cdot u(x))}{\sum_X \exp(\lambda \cdot u(X))} \tag{10.9}$$

Where $\Pr(x)$ is the probability of investing an amount x in the public fund and λ is the weight given to the utility values. If λ is 0 all options have an equal probability, while if λ is equal to infinity the agents choose the option with the highest expected utility.

10.3.7.3 Extraction Decision

Based on the investment decisions of the agents, the actual level of the public infrastructure p can be determined. Now, each agent makes a decision how much to collect, based on the available resource at the turn she can make the decision. Similarly to the investment decisions, the expected utility for each level of collection is determined, and decisions are made from upstream to downstream.

10.3.7.4 Learning

The agents update the expected level of cooperation η_i based on the information they received on the average investments of the other agents. The learning parameter τ_1 defined the speed of learning. If τ_1 is equal to 1, agents do not learn, while if τ_1 is equal to 0 agents assume that the level of cooperation in the nest round is the same as observed in the current round.

$$\eta_i = \eta_i \cdot \tau_1 + (1 - \tau_1) \div \frac{\overline{x}_{-i}}{10} \tag{10.10}$$

Similar for the expected share which upstream agents are expected to extract, we assume that agents update the value of s_i based on the observed share, where τ_2 is a learning rate.

$$s_i = s_i \tau_2 + (1 - \tau_2) \frac{y_{i,t-1}}{p_{t-1}} \tag{10.11}$$

10.4 Overview Model Parameterization

Within the model parameterization framework this is case 9. We do not have a large N to simulate, since we only have five participants in each experiment. The field work does not cover 100 % of the population since we use data from a sample of the populations from a few communities. We have done field work in the community and focus our model on a few behaviors.

M1: To define the model we make use of theory, especially rational choice theory with modifications based on behavioral game theory (Camerer 2003). We assume agents learn, have other regarding preferences, and have a trembling hand when making decisions. These assumptions are confirmed by the interviews and the statistical analysis of the experimental data (Janssen et al. 2012).

M2: Attributes of the agents are based on theory and observations, namely ability to learn, having other regarding preferences. We do not distinguish social-demographic variables since they were not found to have significant impact in the statistical analysis of the experimental analysis. As such the attributes of the agents are derived from the calibration process itself.

M3: The method to collect behavioral data is the performance of field experiments.

M4: Each agent has a unique parameter setting which is the result of parameter distributions. The parameter distributions are the outcome of the calibration process. For simplicities sake we assumed truncated normal distributions. The data did not suggest to explore bimodal or other multi-model distributions. However, we acknowledge that alternative assumptions on agent types could be used to calibrate the model.

M5: The model is not scaled up, although the resulting general model could be used to inform models of irrigation at a watershed level.

10.5 Technical Details

10.5.1 Experimental Data

Before we discuss the parameterization of the model, we first discuss the outcomes of the experiments. A detailed analysis of the experimental results can be found in Janssen et al. (2012). Figure 10.1 shows the average level of contributions to the public fund by all villager and student groups in each round. Figure 10.1 shows that the average investment is around 50 % and reduces only slightly.

In the Fig. 10.2, we report the average contributions to the infrastructure in each of the locations in the watershed. There is no difference in the level of contributions to the infrastructure among the different locations. However, there is a significant inequality of the extraction levels across participants upstream, A and B, and the participants downstream, D and E (Fig. 10.2).

The inequality can also be quantified by gini coefficients. We calculated the gini-coefficients for investments as well as extractions. The average gini coefficient for investments is 0.27 and this level is not changing significantly. The average gini coefficient for extractions is 0.44 and also here this level is not changing significantly.

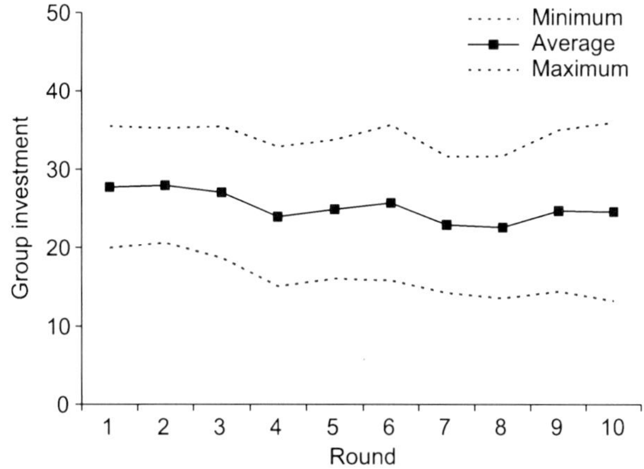

Figure 10.1 The average group investments for 10 rounds. The dotted line is the average ± the standard deviation of the 32 experiments

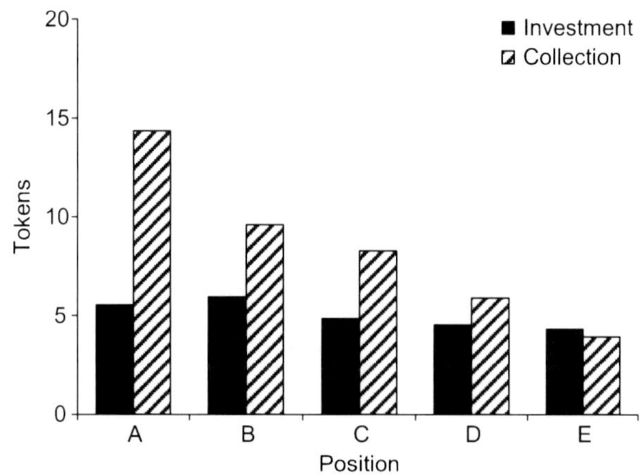

Figure 10.2 The average contribution and collection level per round for each position

Janssen et al. (2012) report detailed statistical analysis which shows that the results are not affected by the type of participants, whether they are irrigators, fishers, or forest appropriators, or students. What is affecting the initial level of investments is the trust in other people of the community. The investment levels in subsequent rounds are affected by the level of inequality of the extractions of the common resource.

10.5.2 Calibration

We calibrated the model on the experimental data. We used the standard genetic algorithm of *BehaviorSearch.org* for the model that is implemented in Netlogo 4.1.3. The model code and documentation can be found at http://www.openabm.org/model/3073/version/1. For the fitness evaluation of each parameter configuration we run the model 100 times 32 runs. For each of the 32 runs we compare the simulated statistics with the actual statistics.

The fit between the model and the data is defined as follows. For each of the metrics included, we calculate the fitness score between 0 and 1, using

$$f_i = 1 - (d_e - d_s)^2 \quad (10.12)$$

where the data of the experiments, d_e, and simulations, d_s, are scaled to values between 0 and 1. Then the fitness values of all five metrics are aggregated to derive the final fitness score used in the calibration. We compare the results of three ways of aggregating the information as spelled out below.

The metrics included to evaluate the performance of the model include:

- Average group level investments in the public infrastructure level over the 10 rounds. To calculate f_1 the simulated average group level investment is denoted by x_{si} and the average group level investment of the data is denoted by $\overline{x}_{di}$. Since the maximum group investment level is 50, the difference is divided by 50 to scale f_1 between 0 and 1.

$$f_1 = 1 - \frac{1}{10} \cdot \sum_{i=1}^{10} \left(\frac{x_{si} - \overline{x}_{di}}{50} \right)^2 \quad (10.13)$$

- The average contribution per position. In the calculation of f_2 the simulated average level of investments per position over 10 rounds is represented as x_{sj}^{pp} and the data as $\overline{x}_{dj}^{pp}$. To scale f_2 between 0 and 1, the difference is divided by 10 since contributions per person are up to 10 tokens.

$$f_2 = 1 - \frac{1}{5} \cdot \sum_{j=1}^{5} \left(\frac{x_{sj}^{pp} - \overline{x}_{dj}^{pp}}{10} \right)^2 \quad (10.14)$$

- The average collection per position. In the calculation of f_3 the simulated average level of collections per position over 10 rounds is represented as y_{sj}^{pp} and the data as $\overline{y}_{dj}^{pp}$. To scale f_3 between 0 and 1, the difference is divided by 100 since the maximum collection a person can make is 100 tokens (the maximum resource).

$$f_3 = 1 - \frac{1}{100} \cdot \sum_{j=1}^{5} \left(\frac{y_{sj}^{pp} - \overline{y}_{dj}^{pp}}{100} \right)^2 \quad (10.15)$$

- The average gini coeffient of contributions. G_s^{con} is the gini coefficient of investment levels in a round average of 10 rounds in each game, averaged over all the simulated games. Similarly, G_d^{con} represented the average gini coefficient per round for all the data.

$$f_4 = 1 - (G_s^{con} - G_d^{con})^2 \tag{10.16}$$

- The average gini coeffient of collected tokens. G_s^{col} is the gini coefficient of harvested tokens in a round average of 10 rounds in each game, averaged over all the simulated games. Similarly, G_d^{col} represented the average gini coefficient per round for all the data.

$$f_5 = 1 - (G_s^{col} - G_d^{col})^2 \tag{10.17}$$

There are different ways to aggregate the individual fits with the indicators. We distinguish three ways, and compare the impacts on the calibration of using the different aggregation methods. The first approach is to multiply the fitness scores of the five indicators.

$$f^{mlt} = f_1 \cdot f_2 \cdot f_3 \cdot f_4 \cdot f_5 \tag{10.18}$$

The second approach is to calculate the average of the fitness scores:

$$f^{avg} = \frac{(f_1 + f_2 + f_3 + f_4 + f_5)}{5} \tag{10.19}$$

and finally, the calibration can be evaluated on the minimum level of all five fitness scores.

$$f^{\min} = \min(f_1, f_2, f_3, f_4, f_5) \tag{10.20}$$

10.5.3 Calibration Process

We first discuss the results of the calibration of the three different fitness functions. Thereafter we perform a sensitivity analysis of the model and present a simpler model that captures the basic results. Our aim is to understand the importance of the basic components of the model and develop a simple model that represents the key empirical features of the model.

When we perform a calibration we run a genetic algorithm with 50 individual randomized starting conditions. We present the parameter values of the 100 best solutions found and the best fitness score. The genetic algorithm, using the standard genetic algorithm from *BehaviorSearch 0.72 (beta)*, has a population size of 50, a mutation rate of 0.01, and a cross over rate of 0.7. We stop the genetic algorithm after 1,000 fitness evaluations. It is important to note here that the fitness is only calculated for new parameter combinations.

Each parameter is distributed using a truncated normal distribution (Table 10.2). The truncation means that drawn parameter values which are not within the eligible range of the parameter are not used, and a new parameter value is generated.

Table 10.3 Fitness scores for the different solutions and for the different aggregation functions

	Nash equilibrium	Cooperative solution	Max f^{mlt}	Max f^{avg}	Max f^{min}
f^{mlt}	0.008	0.008	0.813	0.809	0.800
f^{avg}	0.329	0.427	0.959	0.960	0.957
f^{min}	0.246	0.254	0.926	0.923	0.937

Table 10.4 Parameter distributions for the optimal calibrations for each of the different fitness functions used

	Max multiplier		Max average		Max min	
	Mean	Stdev	Mean	Stdev	Mean	Stdev
α	1.00	0.14	0.96	0.12	0.88	0.05
β	0.52	0.00	0.52	0.00	0.49	0.01
λ	3.10	0.27	3.50	0.38	1.50	0.01
η	0.37	0.00	0.52	0.07	0.33	0.01
τ_1	0.51	0.00	0.56	0.00	0.82	0.04
τ_2	0.71	0.07	0.57	0.00	0.38	0.05

Stdev standard deviation

Table 10.3 shows the fitness scores for the different aggregation metrics and the different solutions. When agents are selfish and rational, and their behavior leads to the Nash equilibrium, the fitness scores are lower compared to the cooperative solution. Hence the behavior of the participants is closer to the cooperative solution compared to the Nash equilibrium. The fitness scores of the results of the calibration are higher. Although the fitness scores are highest for the aggregation function maximized, they are closely together.

Table 10.4 shows the optimal parameter distributions for the different calibration functions. We use the data from the best solution of the 50 runs of the genetic algorithm.

The agents have a strong aversion to exploit others (α) and a strong tendency to altruism (β). Agents are expecting an initial level of cooperation of about 50 % from the fellow participants. Learning is slow (τ value are often higher than 0.50).

When we plot the data with the simulations results of the calibrated models, we see that the three different calibrations do not lead to main differences in the results (Figs. 10.3, 10.4, 10.5, 10.6). The patterns in the simulated data are smoother than the actual data. This is not strange since the simulated data is smoothed over 100 runs. We also see that the distribution of the collected tokens is more spread in the actual data than the simulated data (Fig. 10.5).

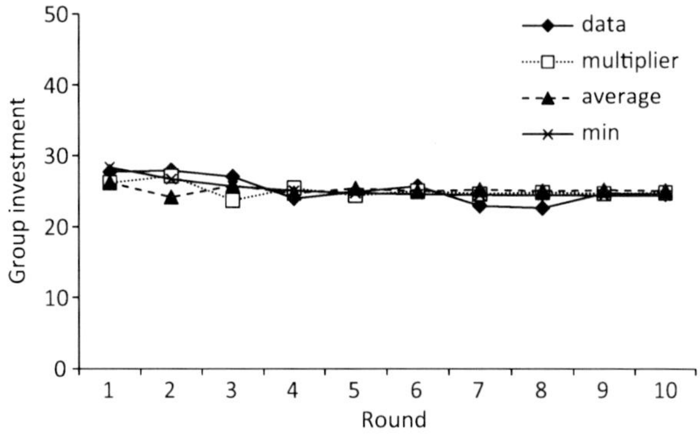

Figure 10.3 Average group investment levels per rounds for the data and the optimal calibrations using three different fitness score functions

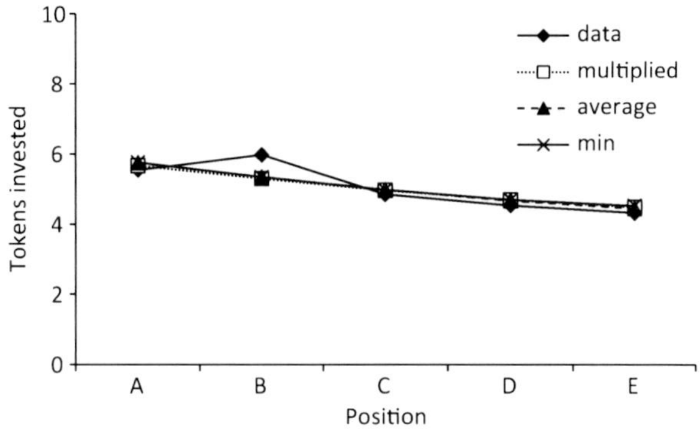

Figure 10.4 Average investment level per position for the data and the optimal calibrations using three different fitness score functions

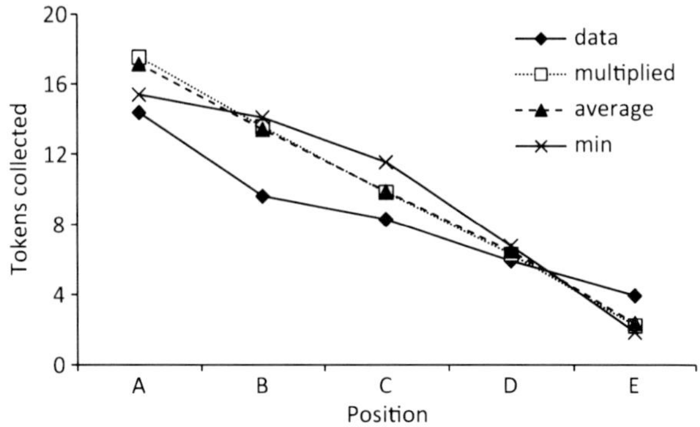

Figure 10.5 Average harvesting levels per position for the data and the optimal calibrations using three different fitness score functions

10.5.4 Sensitivity Analysis

A sensitivity analysis was performed to test the importance of particular assumptions of the model. For five cases we held parameters constant and analyzed the consequences for changes the mean of one of two parameters. For this analysis

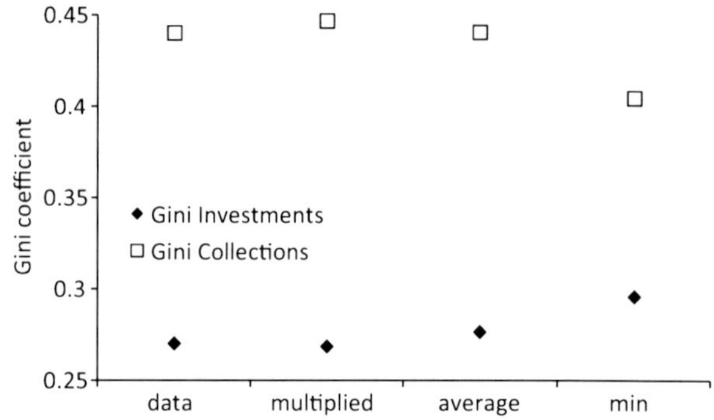

Figure 10.6 Average gini scores for investments and harvesting rate for the data and the optimal calibrations using three different fitness score functions

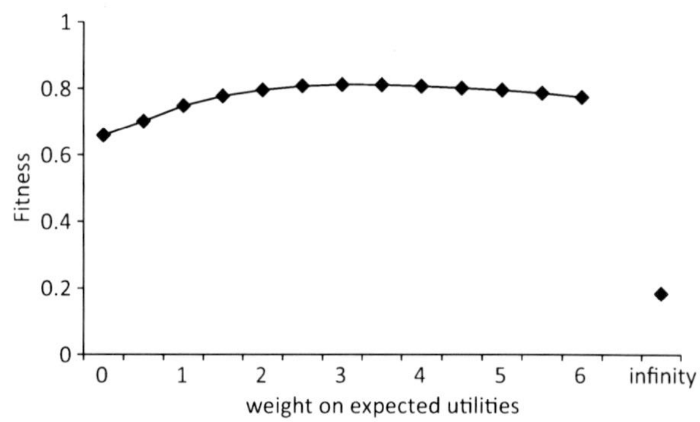

Figure 10.7 The effect of different values of λ on the fitness of the model with the data

we uses only one fitness function, namely f^{mlt}, since the analysis above shows the robustness of the results for the different fitness functions.

The first case was the assumption of probabilistic choice (Fig. 10.7). If agents always choose the option leading to the highest utility, the fitness score is dramatically reduced (from 0.8 to 0.2). With a probabilistic choice, we see that the model is not very sensitive to modest changes in the values of λ. If λ is close to zero and agents make random decisions, we see that the fit is significantly reduced. Note that this means that a null model that makes random decisions from the options available (fitness = 0.66) is doing a better job than the theoretical solution of the Nash equilibrium and the cooperative solution who have a fitness value of 0.008.

If agents do not learn ($\tau_1 = 1$), there is only a slight reduction of the performance. The performance of the model is insensitive to the learning rate of the share that upstream participants have extracted. The model is sensitive to changes in the learning rate of cooperation which affect the expectations of investments. If agents expect that observations in the current round will be the best estimate for the next the performance of the model drops. Hence agents will have a learning rate that takes into account observations of the last round and the long term trend. This avoid agents will overreact to impulsive decisions made by others. In fact agents can forgive a selfish decision by not adjusting the cooperating expectations immediately in the next round (Fig. 10.8).

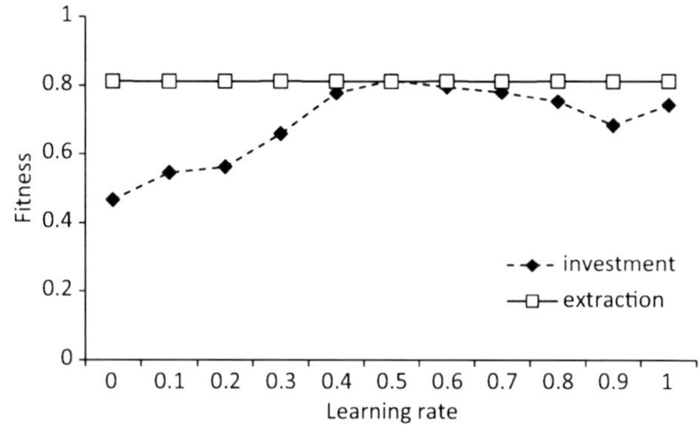

Figure 10.8 The effect of different learning rates τ_1 and τ_2

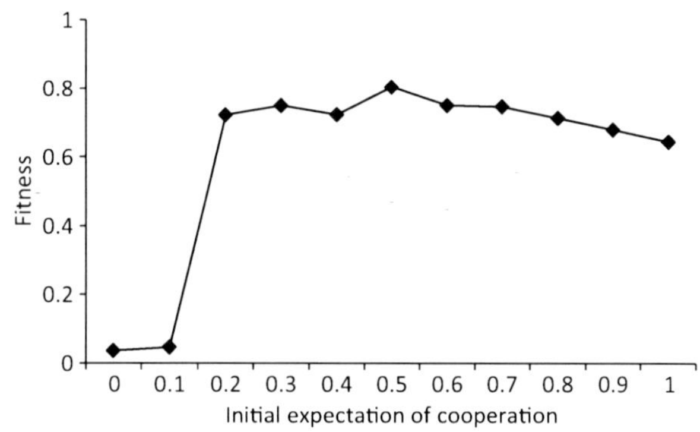

Figure 10.9 The effect of different values of the expected level of cooperation by others

When agents assume initially that other agents do not cooperate, the performance of the model drops to almost 0. Agents will not cooperate and the investment levels will be close to 0. The performance of the calibration is not much affected by the assumption of the initial expectation what other agents do is the expected level of cooperation is around 50% (Fig. 10.9).

Finally, if agents are assumed to be selfish and α and β are equal to 0 the performance of the model is poor. The model performance starts to improve if β is 0.5 or 0.6 and α is 0.5 or higher. The sensitivity analysis shows the importance of including other regarding preferences within the model formulation (Fig. 10.10).

10.5.5 Simple Model

Based on the sensitivity analysis we define a simplified model. We assume τ_1 is equal to 0.5, τ_2 is equal to 0, and λ equal to 3, and η is equal to 0.5. We keep the other regarding preferences as found in the calibration of the model. The fitness score is 0.801 and only slightly lower than the best score of 0.813.

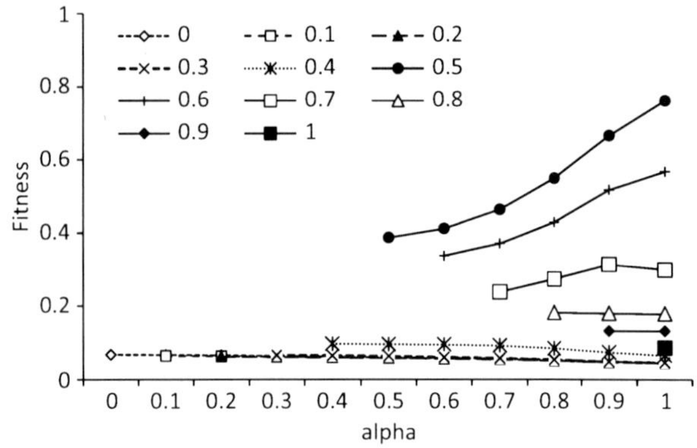

Figure 10.10 The effect of different values of α and β. The horizontal axis refers to different values of α, while the different lines refer to different values of β

10.6 Lessons Learned

In this chapter, a detailed calibration exercise is discussed of an agent-based model on field experimental data. The model was constructed to represent a general model of decision making in social dilemmas. It captures insights from behavioral games theory, namely assumptions on other-regarding preferences, and learning. The model formulation was also affected by the insights from the statistical analysis of the particular field experiments.

The model did not aim to capture irrigation systems within a particular context, but decision making in a general context that resembles dilemmas irrigators have. With that in mind we use data from six villages and two student populations to create a data set of 32 groups. The aim of the model was to capture the general patterns coming out of these 32 groups.

The traditional model of rational selfish actors cannot explain the data, nor do the data support a cooperative solution. There is not a clear candidate model of the observations. Hence our model analysis contributes to the development of an alternative theory of collective action based on experimental data.

A challenge is to define the performance of a model. Different indicators provide information on the dynamics within the field experiments. These indicators capture distributions and averages at the group and individual level. We define five different patterns to capture the dynamics of the system.

Grimm et al. (2005) define pattern oriented modeling to develop models that capture as many relevant patterns as possible. Typical pattern oriented modeling exercises explore the parameter space to find parameter settings that capture the patterns within the uncertainty ranges. This has also been performed with experimental data (Janssen et al. 2009).

In this paper, we decided to evaluate the model on the different patterns using optimization. We maximize the fit between data and simulations. Since there are different ways to aggregate information on performance on five different patterns, we tested the effects of these different fitness functions. The results are robust to the different fitness functions.

The model was implemented in Netlogo and we could make use of the BehaviorSearch tool developed for Netlogo. This enables us to easily apply optimization to agent-based models. BehaviorSearch include different optimization algorithms (random search, genetic algorithm, and hill climbing). We made use of genetic algorithms because of its proven performance on complex optimization problems. Since genetic algorithms do not always lead to the same results, we used 50 different runs for each optimization task.

In our sensitivity analysis, we found particular assumptions of the model critical for the performance of the model. One critical assumption is the trembling hand, meaning that agents do not choose the option with highest utility. Another critical assumption is that agents are not selfish. Finally it is critical to assume that agents assume other agents have a modest level of cooperation. Based on our sensitivity analysis we could propose a simplified version of the model, which might be useful for other studies on irrigation systems and collective action.

Reflecting on the calibration analysis, we found a systematic sensitivity analysis helpful to understand which aspects of the model are critical for the model performance. As such we could propose a simplified model. The goal of this exercise is therefore not to find the specific parameter values, but to contribute to an alternative model of decision making in collective action situations. Our analysis confirms the critical aspects of other regarding preferences and the trembling hand in decision making.

References

Camerer CF (2003) Behavioral game theory: experiments in strategic interaction. Princeton University Press, Princeton

Charness G, Rabin M (2002) Understanding social preferences with simple tests. Quart J Econ 117:817–869

Fehr E, Schmidt KM (1999) A theory of fairness, competition, and cooperation. Quart J Econ 114(3):817–868

Grimm V, Revilla E, Berger U, Jeltsch F, Mooij WM, Railsback SF, Thulke HH, Weiner J, Wiegand T, DeAngelis DL (2005) Pattern-oriented modeling of agent-based complex systems: lessons from ecology. Science 310:987–991

Hunt RC (1988) Size and structure of authority in canal irrigation systems. J Anthropol Res 44(4):335–355

Janssen MA, Radtke NP, Lee A (2009) Pattern-oriented modeling of commons dilemma experiments. Adapt Behav 17:508–523

Janssen MA, Bousquet F, Cardenas JC, Castillo D, Worrapimphong K (2012) Field experiments of irrigation dilemmas. Agricultural Systems 109:65–75

Lansing JS (1991) Priests and programmers: technologies of power in the engineered landscape of Bali. Princeton University Press, Princeton

Ostrom E (1992) Crafting institutions for self-governing irrigation systems. ICS Press, San Francisco, CA

Wittfogel KA (1957) Oriental despotism: a comparative study of total power. Yale University Press, New Haven, CT

Chapter 11
Companion Modelling with Rice Farmers to Characterise and Parameterise an Agent-Based Model on the Land/Water Use and Labour Migration in Northeast Thailand

C. Le Page, W. Naivinit, G. Trébuil and N. Gajaseni

11.1 Introduction

Rainfed lowland rice (RLR) is the dominant type of land use in Northeast Thailand, a cradle of this crucial staple crop in continental Southeast Asia. Low water-holding and infertile coarse-textured soils, added to erratic rainfall distribution lead to low paddy yields of the single wet season crop cycle, and very low per capita farm incomes. Therefore, to improve their livelihoods, young members of the resource-poor smallholdings have long been migrating to urban areas and abroad, on a seasonally or recently more permanent basis. This can cause labour scarcity at the household and community levels during the periods of peak labour demand at RLR transplanting and harvest.

Past research and development efforts focused mainly on improving the drought tolerance of rice varieties (Jongdee et al. 2006) and focused on the hydrological functioning at the paddy field level (Trébuil et al. 1998). How agricultural water use is decided and implemented at the whole farm level remains largely unknown. To better understand the interactions between the water-resource and water-use dynamics in the RLR ecosystem, a first agent-based simulation tool was developed

C. LePage (✉) · G. Trébuil
UPR Green, Cirad, 34000 Montpellier, France
e-mail: christophe.le_page@cirad.fr

W. Naivinit
Ubon Ratchathani University,
Warinchamrap, 34190 Ubon Ratchathani, Thailand
e-mail: wnaivinit@gmail.com

N. Gajaseni
Chulalongkorn University, Bangkok, Thailand
e-mail: gnantana@chula.ac.th

G. Trébuil
e-mail: guy.trebuil@cirad.fr

A. Smajgl, O. Barreteau (eds.), *Empirical Agent-Based Modelling – Challenges and Solutions*, DOI 10.1007/978-1-4614-6134-0_11,

(Lacombe and Naivinit 2005). Based on expert knowledge, this model provided a detailed representation of the hydrological processes at work in a small RLR catchment. It was constructed in 6 months by a French master student and his supervisor, and it was not possible to present its final version to the local rice farmers to assess the relationships between simulation results and their long empirical experience. In a following phase, to tackle the complexity of the intertwined social and hydrological dynamics, it was decided to represent the interactions between RLR farming, water availability, and labour migrations into a second agent-based model (ABM). This time, a truly collaborative modelling process was conducted by Warong Naivinit, a Thai PhD candidate who collaborated in designing the first expert-based hydrological simulation model. Under his guidance, a group of local farmers from the *Ban Mak Mai* village in Northeast Thailand, representing the diversity of local RLR growers, engaged in a four year long collaborative modelling process. At the end, four collaborative farmers presented the final version of the model to an academic audience composed of tens of master students and their lecturers at the Faculty of Agriculture of the regional university.

11.2 Overview, Design and Details of the Ban Mak Mai Agent-Based Model

Following the ODD protocol (Grimm et al. 2006, 2010), a shortened description of the model is provided below. The full details are available in Naivinit et al. (2010).

11.2.1 Purpose of the BMM Model

The BMM model is a communication tool that is used by scientists and local RLR farmers to exchange and integrate knowledge about the interactions between land and water use and labour migration in the RLR environment of lower Northeast Thailand.

11.2.2 State Variables and Scales

In the BMM model, individuals (members) belong to households, households belong to a village. Members have specific age, gender, marital status and migration experience. The age of a member influences its worker status and role (dependant, farm worker or migrant) while gender, marital status and migration experience influence an individual's decision to migrate or not.

All main RLR-producing activities are decided at the household level by considering the need for and the availability of water whenever it is relevant (there are thresholds of daily rainfall to start the nursery bed, to start transplanting and to pause harvesting).

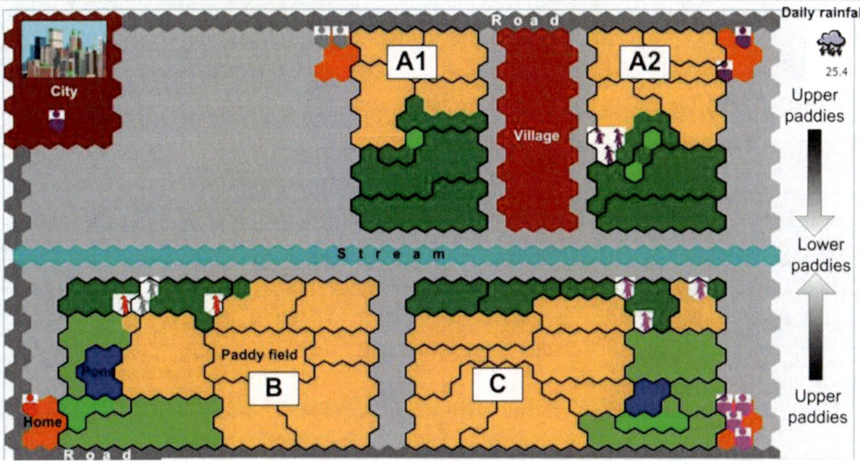

Figure 11.1 Spatial setting and visualization of agents' activities

The village level provides a registration desk where all potential farm workers are listed for hiring. The daily wages for transplanting and harvesting rice are also defined at this level.

In the BMM model, the two most common aromatic photosensitive late-maturing rice varieties are planted: the glutinous (RD6 variety) type for self-consumption, and the non-glutinous (KDML105 variety) rice for sale. The key dates and durations related to the successive phases of the RLR-based cropping system (seedling stage, transplanting, harvesting) are state variables of these two rice classes.

In the BMM model, water tanks are either paddy field ponding tanks (20 cm deep) or farm pond storage tanks (3 m deep). The water level in the tanks is updated on a daily basis depending on the balance between rainfall and evaporation read from external data files. When the water level exceeds the height of a water tank, the overflowing water is shared with the lower level neighbouring water tanks (runoff). Additionally, an estimated constant volume (10 mm per day) is subtracted from the ponding tanks to account for water used by the soil-plant system. Therefore, compared to the first model mentioned earlier (Lacombe and Naivinit 2005), the hydrological processes were substantially simplified. This was decided in consultation with the collaborative growers.

The spatial resolution was set to 0.04 ha (1 "ngan", a traditional Thai unit) and the main spatial interface (see Fig. 11.1) was set to represent in a minimalist way a typical portion of a RLR ecosystem at a village level made of a few farms varying in size. Farms are made of paddy fields defined as aggregates of 7–24 contiguous cells (0.28 ha to 0.96 ha).

A daily time step was chosen because participating rice farmers adjust their cropping decisions according to climatic conditions on a daily basis. However, whereas water levels are updated in the tanks daily, farming and migration decisions are made at specific times of the cropping calendar, some of them being related to the water dynamics (for instance occurrences of water stress during the nursery stage). The time horizon was set to 5 years.

11.2.3 Process Overview and Scheduling

A simple hydro-climatic process aggregating rainfall, evaporation, run-off and soil-plant consumption is run on a daily basis to update the water levels in all water storage tanks and to determine water availability for rice cropping. Farming and migration decisions are made at specific times, some of them being related to the water dynamics (for instance occurrences of water stress during the nursery stage). The key successive farming activities are as follows: establishment of RLR nurseries and production of seedlings, transplanting, and harvesting. After RLR harvest, each household computes the economic results of the rice season. This updated household income and the presence of dependants in the household is taken into account when each member makes migration decisions at the beginning of the dry season.

The comprehensive list of the model parameters and their values is presented in Naivinit et al. (2010).

11.2.4 Design Concepts

The BMM model is deterministic. The absence of randomness was crucial for a model to be investigated by farmers who are requested to assess whether the outputs of the simulation are meaningful or not. For them to successfully engage in critical thinking by looking for rational explanations of the simulation results, the variability inherent to the randomness would have been a source of confusion.

Three aggregated social levels are explicitly represented: individual, household, and village. Household agents adapt to labour constraints and can hire extra farm workers at transplanting and harvesting stages if needed (they are able to anticipate the need). A list of farm workers available for hire is updated at the village level, and made accessible to all household agents.

The observation of a simulation was tailored for farmers to easily observe and understand what was happening. The main visualization artefact provides a schematic representation of the main features directly related to the question addressed by the model (see Fig. 11.1). It was designed in such a way they could easily relate the computerized interfaces to the previous stages of the modelling process. Rice was represented on paddy fields with a range of green colours corresponding to the variety and to the stage of the crop cycle.

Household members were depicted in specific locations depending on their role: children (below 15) and elders (over 65) are dependants staying at home (see the orange blocks beside paddy fields in Fig. 11.1) while others can be migrating (see the city image at the top left corner in Fig. 11.1) or working in paddy fields. A place was also set to represent the village, where active and non-migrating members whose paddy fields are not requesting any work, and who are not hired by other households, are located. In this virtual environment, neither the relative sizes of the four households, the village and the city nor the distances among them are meant to be related to the spatial resolution set to represent the paddy fields. As a result, the representation of the environment in the *Ban Mak Mai* ABM is quite abstract. Nevertheless, it also allows the farmers to conveniently get the whole picture of the evolution of the system during a simulation.

11.2.5 Initialization

The design of the initial situation was set according to the typology of farms found in that region: two small (3.6 ha) farms without pond (A1 and A2) and two large (7 ha) farms with a pond (B and C) were schematically represented (see Fig. 11.1). To represent upper and lower paddies, a stream was set in the middle of the virtual environment and a regular gradient of cell elevation was set to create a moderate slope typical of the local undulating small catchments. The locations of rice nurseries (the smallest paddy field made of 3–5 light green hexagons in Fig. 11.1) were given by farmers: as in reality, they are neither far from their house nor from the pond (in case it exists), and higher places are better for water control. In case there is no pond (farms A1 and A2), the nursery was located in the middle of the farm to minimize the bulky transport of seedlings.

The characteristics of individuals from each household were chosen to account realistically for the heterogeneity of family members in the Northeastern Thailand region (see Table 11.1).

11.2.6 Input Data

Daily rainfall and potential evapotranspiration (PET) data used in the model to create the climatic conditions affecting farmer agents' decisions were obtained from the nearby regional meteorological centre located in Ubon Ratchathani Province. The same set of 5 years (1991–1995) was used for all simulation experiments. The indication of daily rainfall (see top right corner in Fig. 11.1) was used in combination with a climatic timetable recording weekly conditions and displaying also the dates of key ceremonies in the traditional lunar calendar used by local villagers, such as the Thai New Year in mid-April and the Royal Ploughing Ceremony in early May (see Fig. 11.2). These milestones helped the farmers to understand the chronology of a simulation and decision-making for RLR production.

11.2.7 Submodels

The details of the submodels are given in Naivinit et al. (2010).

Table 11.1 Characteristics of individuals from each of the four households at initialization

Household	Gender	Age	Marital status	Migration experience
A1	Male	55	Married	Yes
	Female	55	Married	No
	Female	30	Married	Yes
	Male	25	Single	Yes
	Female	10	Single	No
	Male	8	Single	No
A2	Male	55	Married	Yes
	Female	52	Married	Yes
	Female	32	Married	Yes
	Male	29	Married	Yes
	Female	10	Single	No
	Female	6	Single	No
B	Male	50	Married	Yes
	Female	45	Married	No
	Male	5	Single	No
C	Male	50	Married	No
	Female	45	Married	No
	Male	30	Married	Yes
	Female	14	Single	No
	Male	12	Single	No
	Male	5	Single	No
	Female	2	Single	No

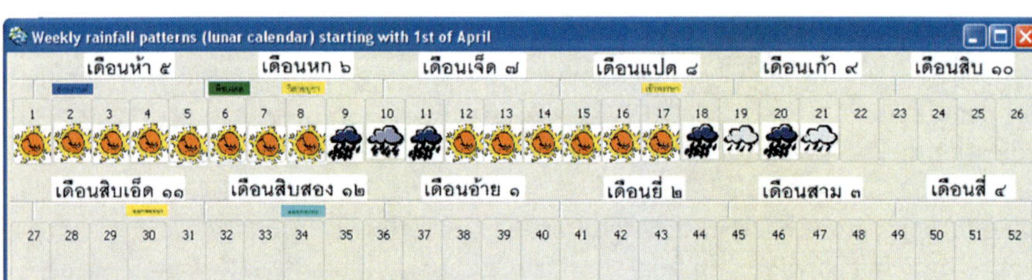

Figure 11.2 Interface displaying weekly rainfall patterns and timing of key traditional ceremonies according to the lunar calendar

11.3 Characterisation and Parameterisation Framework and Specific Sequence of Activities

In this section, we analyze how the *Ban Mak Mai* modelling process can be referred to the characterisation and parameterisation (CAP) framework detailed in Chap. 1 and also to the parameterisation sequences described in the paper by Smajgl et al. (2011).

11.3.1 A Preliminary Expert-Based Characterisation Method

The main model characterisation method (M1), prior to the co-design process, was an agrarian system analysis of the Lam Dome Yai watershed in Southern Ubon Ratchathani Province. The Agrarian Systems theoretical framework focuses on the differentiation processes among resource users in farming communities and their subsequent differences in personal interest and concerns. Three complementary methods are implemented to implement such analysis: an agro-ecological zonation of the area, an historical profile of changes in agricultural activities to understand the origins of the present diversity of farming situations, and, based on the outputs from the previous methods, an analysis of the functioning of the diverse types of farms leading to the construction of a farmer typology based on their assets, socio-economic objectives and strategic decision-making rules in farming (Trébuil and Dufumier 1993). The recent changes in the main interacting socio-economic and agro-ecological dynamics of the system were revealed to understand the historical processes of socio-economic differentiation among the local farming households. Currently, three main types of farming households were classified according to their agro-ecological constraints and opportunities and socio-economic strategies. Type A small-holding farmers was identified as the dominant type in the study area. These tiny holdings play an important role in supplying labour to the community because their land per labour ratio is low. Type A farmers often migrate seasonally. Larger farming units belong to type B and C farmers where labour shortage can be a constraint, with type C farmers having less labour constraints since they are either more mechanized or because their household size is larger. Nevertheless, these farm types, in particular type B, play a major role in employing hired labour during the periods of peak labour demand, particularly at rice transplanting and harvest. The migration pattern for type B and C farmers is more permanent.

11.3.2 Role-Playing Games to Co-Design the Conceptual Agent-Based Model

Within the area of the Lame Dome Yai watershed, the *Ban Mak Mai* village was selected, as a typical regional RLR-based agro-ecological system with a diversity of farming households. This village was chosen because its farming systems were

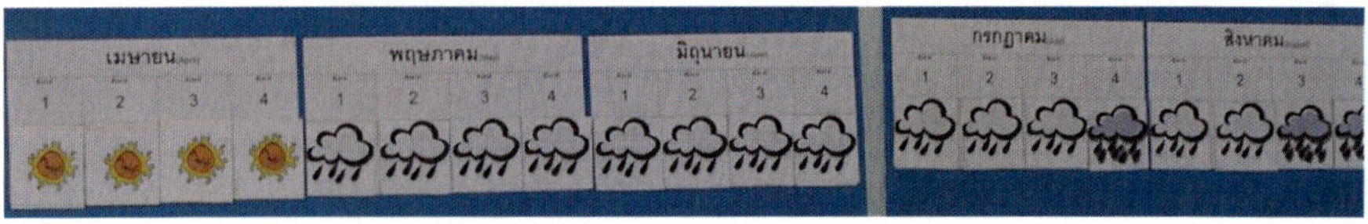

Figure 11.3 Climatic cards drawn during role-playing game sessions

studied a decade before and the results from that previous analysis were useful to document recent changes in the agricultural system. A farm survey based on repeated individual interviews with a sample of RLR growers was conducted in the village to document decision-making regarding land/water use, RLR production and migration practices across the most diverse range of farming households (M2 and M3 methods). Based on this information, 11 farming households representing the diversity of RLR farming situations in the village and ranging from small farms (average size of 3.2 ha), to larger holdings (average size of 7.2 ha) were recruited to reflect the farm typology defined through the M1 method described above. 8 out of 11 participating households belong to the type A, two belong to type B and one to type C. The husbands and wives from each selected household were involved in the co-design of the conceptual agent-based model (both M2 and M3 methods) through a series of modelling field workshops based on role-playing games and held in their village. The workshop participants received a daily small compensation in cash corresponding to the local cost of hiring a farm worker.

During these role-playing game sessions, farmers were free to "play" their singular situation for them to become familiar with the stylization of the reality that was introduced in the game and the conceptual model used to build it. When requested to indicate the situation they would play in terms of planted areas for the two selected rice varieties, the presence of a pond and the composition of the household (gender, age, marital status, migration experience of each family member), all of them decided to stick to their own situation and to replicate what they were used to do in their actual circumstances. Time was set to run, week after week, from April 1 with the drawing of climatic cards (indicating dry weather or light or heavy rain) that were progressively placed on a board until the last plot of rice was harvested (see Fig. 11.3).

The starting date was set in early April as it was clearly ahead of the traditional beginning of the rice-growing season marked by the Royal Ploughing Ceremony taking place in early May, a key milestone for farmers to start their RLR crop cycle.

As time elapsed until the end of the rice cropping season, farmers had to fill a form for the three successive stages of the rice-growing cycle (nursery establishment, transplanting, harvesting) to indicate how they coped with the specific constraints related to water and labour availability. Figure 11.4 shows an example at harvesting stage. The two lines at the top of the decision sheet were used to tick the weeks when farmers would be busy harvesting the two rice varieties. Stickers of a specific colour per household and representing individuals in the household had to be assigned to the following categories: dependant household member, paddy field worker (household member or hired labour), migrating household member.

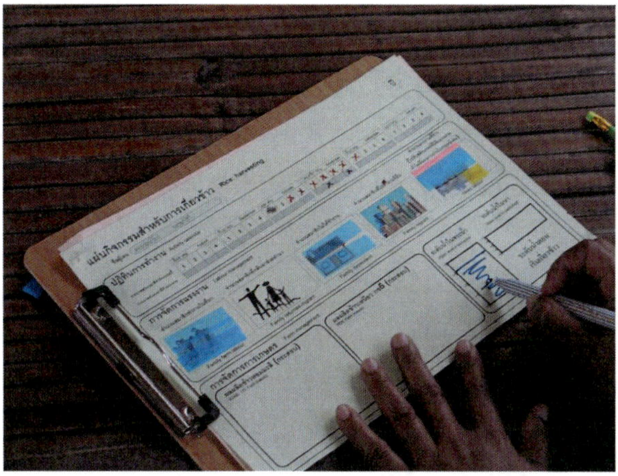

Figure 11.4 An example of the decision-sheet filled by farmers during a role-playing game session

Additionally, to get an idea of the farmers' representation of water availability, water levels in ponds (if any) and in paddy fields had to be drawn.

The role-playing game sessions were important in the overall process as these first steps of the modelling process allowed to instil positive group dynamics and to build trust between the research team and the participating farmers. These sessions were terminated by identifying collectively unclear points to be investigated in the next field workshop. Furthermore, by "replaying" individual decisions during the collective debriefing, it became possible to seek out elements that were common to several participants, paving the way towards the identification and acceptance of the basic concepts on which the *Ban Mak Mai* model would be founded.

11.3.3 Additional Workshops Based on UML Diagrams to Focus on Some Behavioural Components

All the rule-based algorithms related to rice cropping activities (establishment and maintenance of nursery beds, transplanting and harvesting) were co-designed with a group of (more motivated) farmers which size (6–10) was reduced compared to the group of farmers (15–18) who participated to the role-playing game sessions. Looking for a consensus in identifying and calibrating the criteria used to make particular decisions, the discussions were organized through the design of UML activity diagrams. Figure 11.5 shows an example of such an outcome regarding the decision to establish the nursery bed.

The actual date of the Royal Ploughing Ceremony held in Bangkok is usually in early May. Every year the date of this ritual varies as it is determined by Brahmin astrologers. The date for 2009 (May 11th) was used to set the default value of a key-parameter of the model: the earliest date to establish a nursery bed. To strictly follow this rule can be seen as a way to cope with climatic uncertainty by minimizing the risk to encounter long periods of successive dry days, which for farmers without irrigation water, would lead to the failure of their seedlings in the nursery.

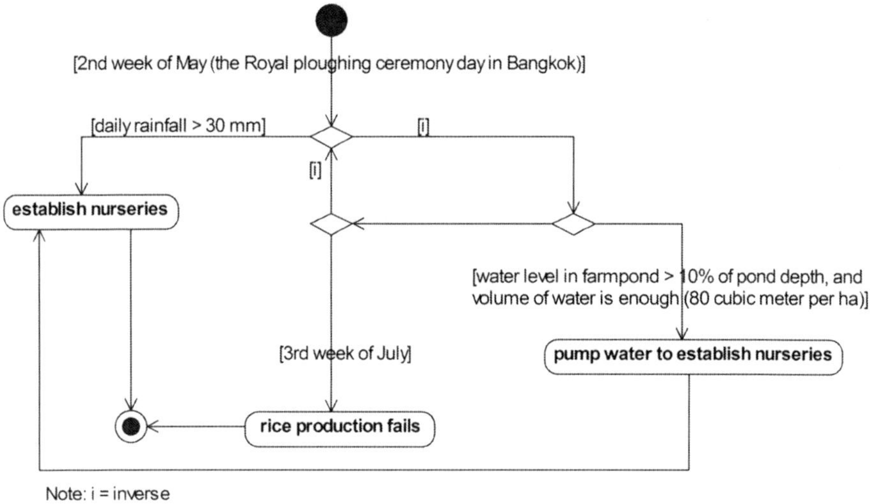

Figure 11.5 UML activity diagram for the establishment of rice nurseries. (From Naivinit et al. 2010)

11.3.4 Early and Iterative Use of the Simulation Model to Fine Tune the Conceptual Model and to Identify Scenarios

Running first versions of the prototype ABM was also a mean to elicit attribute date (M2) and behavioural data (M3). We were in the case of an early and iterative field-based use of simulation with the specific purpose to calibrate the model and to identify scenarios of interest to be explored and collectively assessed later on with the model users. The first version of the *Ban Mak Mai* ABM was introduced to the participating farmers through the simulation of the baseline scenario corresponding to the situation described above in the ODD protocol. We wanted to check whether the participating farmers would react to potentially inconsistent or hard-to-explain results. Facing incongruities, would they be stimulated to propose possible explanations, corrections or improvements? Showcasing the "business as usual" scenario enabled to fine tune the means of observation: a list of indicators was included to make an economic assessment of the cropping season. The farmers were requested to suggest scenarios of interest to them to be explored with the model. They came out with two proposals. The first one was related to labour availability with the introduction of cheap foreign labourers from neighbouring Lao PDR and Cambodia during the RLR transplanting and harvesting phases. Its simulation induced a shortfall for smallholders A1 and A2, while farms B and C achieved higher incomes. During the discussion, farmers mentioned that cheaper wages were not the only criterion to be taken into account, and that recognised farming skills were also essential. The second scenario was related to water availability: what would happen if all farms had enough irrigation water (in the context of new promises made by the authorities to develop irrigation infrastructures in the area)? Unanticipated outputs triggered an interesting discussion about the consequences of a synchronisation of the RLR farming calendar, made possible by water control, on the availability of labour. These two scenarios were combined to conduct a laboratory-based simulation experiment and to perform an in-depth sensitivity analysis (Naivinit 2009).

11.4 Overview of the Overall Sequence

The successive versions of RPGs and ABMs used in combination supported each other in the system analysis and helped to gradually improve the shared representation of the system (see Fig. 11.6). The changes made in the model, moving from somewhat concrete (realistic) features to more abstract ones, reflect the evolution of the conceptual model and its improvement to better facilitate knowledge exchange, which is the main objective of such a companion modelling approach (Etienne 2011).

Each workshop focused on a particular aspect. For instance, during the workshop based on RPG3, the emphasis was on daily rainfall thresholds related to specific farming decisions (to start establishing the nursery bed, start transplanting, or to suspend harvesting). During the workshop based on ABM2, the labour constraints related to transplanting and harvesting were specified. Hence, the overall process looks more like a continuous co-design one than a process of revising a first version of a model.

11.5 Lessons Learned

The ultimate goal of a participatory modelling process is not necessarily to straightforwardly support decision-making, policy, regulation or management by identifying and clarifying the impacts of solutions to a given problem. Viewed as a collaborative learning process, it may aim at enhancing the stakeholders knowledge and understanding of a system and its dynamics under various conditions (Voinov and Bousquet 2010). Walker and his colleagues (Walker et al. 2002) proposed to put into place the stakeholder-led development of a conceptual model of the system as the first step of a framework aiming at fostering the sustainability of this system. Later on, it can be used in exploring ways to maintain the system functionality when it is perturbed, or to maintain the elements needed to renew or reorganize it if a profound perturbation radically alters its structure. Using a model, which for a computer model means playing with it by running simulation experiments, represents only one side of the investigation process. Moving often and early back and forth from the conceptualization to the simulation enable the participating stakeholders to easily relate what is exhibited by the simulation to the structure of the model. Furthermore, this is also raising the stakeholders' awareness about the underlying model

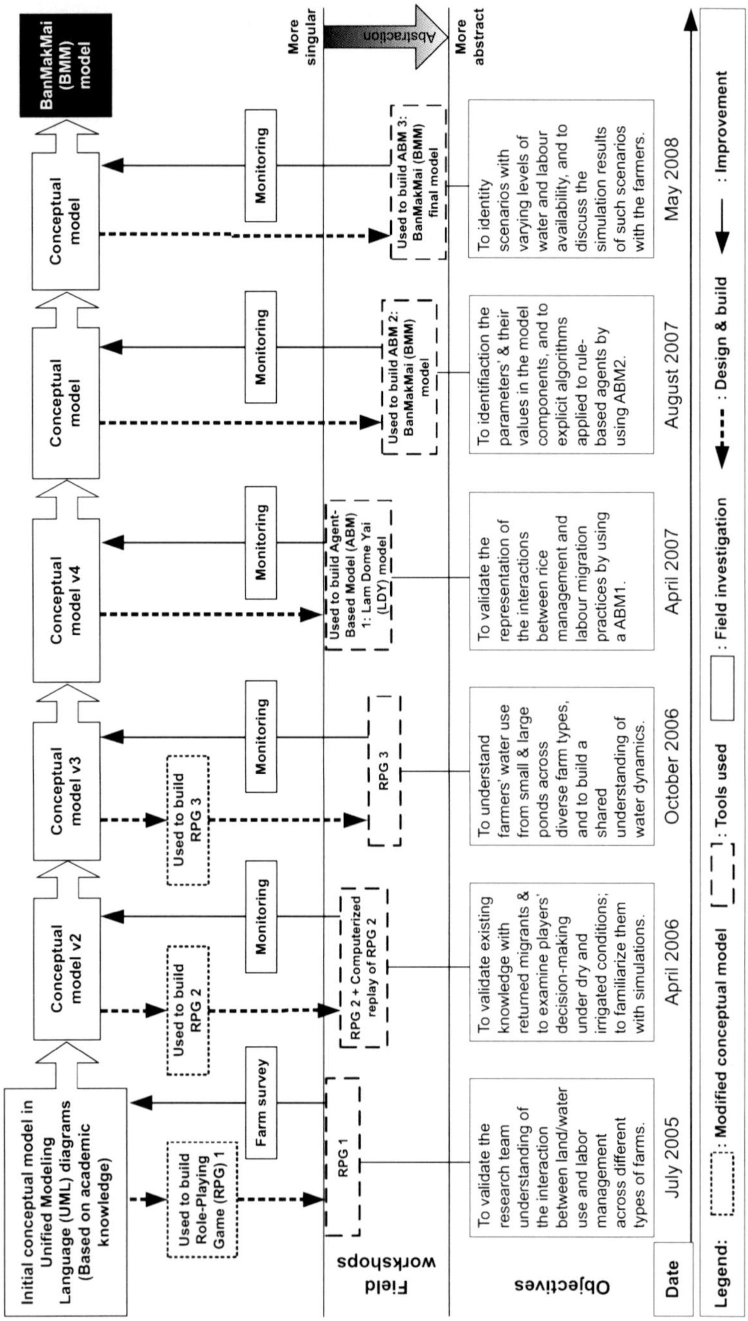

Figure 11.6 overview of the overall modelling process in Ban Mak Mai. (Adapted from Naivinit et al. 2010)

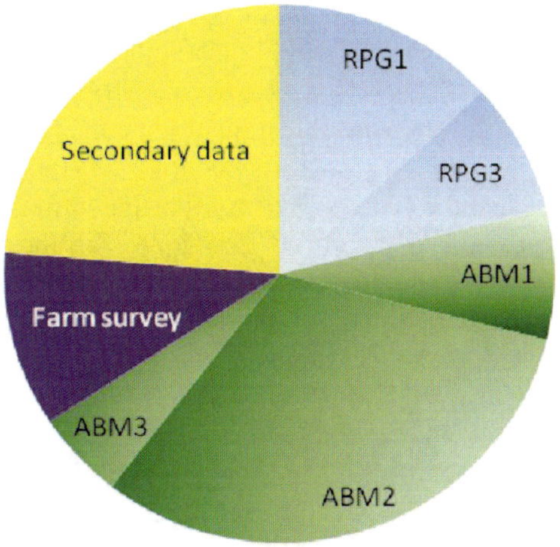

Figure 11.7 Relative importance of several sources of information used to specify the parameters of the BMM model. (Based on Table 1 from Naivinit et al. 2010) Nota: RPG 1, 2, 3 and ABM 1, 2, 3 are successive versions of the role-playing game and agent-based simulation tool respectively

assumptions and simplifications. When invited to challenge these assumptions or to suggest structural modifications, they progressively gain a genuine understanding of what the model is made of and can do for them.

In the case of the *Ban Mak Mai* model, the process was long (4 years). Time was needed because the march towards abstraction is long as the process is unnatural to any human being: it is easier to roll out cognitive processes rather than looking back at them to analyse and dissect them. With local stakeholders, role-playing game sessions are particularly useful to initiate this process of abstraction by collectively setting the conceptual basis of the model (Naivinit 2009, Chap. 8). The progressive refinement of the concepts took place through the succession of field workshops (see Fig. 11.6). The artefact was adjusted each time to the specific purpose of each workshop, leading to the production of a lineage of models, each of them having a different focus. Nevertheless, nearly half of the parameters of the model final version were set during workshops based on the autonomous agent-based model (see Fig. 11.7).

When modelling with local stakeholders, the artefacts used to support the workshops must be stimulating to overcome the participants' reluctance to engage in the arduous process of abstraction, especially when they are rather poor villagers living in a remote corner of the country like in this case. Being able to revisit algorithms and straight away to simulate them with the simulation model modified on the fly is very efficient to raise and maintain their interest (in this case, the collaborative farmers remained the same throughout the whole process). Tools like the executable UML diagrams proposed by Bommel et al. (2011) and recently tested with Uruguayan farmers seem to be promising ones.

At completion of this *Ban Mak Mai* companion modelling process, a special seminar was organized on 18 October 2008 at the regional Faculty of Agriculture of Ubon Ratchathani University during which four collaborative farmers presented "their" model on rice farming and labour migration in front of seventy graduate students and the faculty staff (a presentation of this seminar can be seen at http://www.ecole-commod.sc.chula.ac.th/pn25/index.php?option=com_content&task=view&id=70&Itemid=70). Following a presentation of the model and its use to simulate scenarios, a long question & answers session took place during which the rice growers were able to justify the many choices they made in the construction of their simulation tool. For example, the students and their lecturers were surprised to see that the ABM does not offer the option of establishing the RLR crop by direct seeding (a popular new technique with local extension workers). The farmers explained that, under their village conditions, they still have enough time and labour available to transplant their rice seedlings manually and gradually depending on the arrival of the wet monsoon rains in June–July. They also underlined the fact that transplanting is the best way to avoid weed infestations, a major limiting factor of the yield in direct seeded rice. This session was a convincing demonstration of the farmers' confidence in their knowledge and the suitability of the simulation tool. It was also a way to show to the graduate students and the faculty staff how far collaborative modelling and simulation activities could change the relationships between farmers, extension workers, and on-farm researchers in favour of the co-design and testing of agricultural innovations and infrastructures.

References

Bommel P, Dieguez F, Morales H, Bartaburu D, Duarte E, Montes E, Pereira M, Corral J (2011) One more step towards participatory modeling. Involving local stakeholders in designing scientific models for participative foresight studies. 7th European Social Simulation Association Conference, September 19–23 2011, Montpellier, France

Etienne M (ed) (2011) Companion modelling. A participatory approach to support sustainable development. Editions Quae, Versailles, p 347

Grimm V, Berger U, Bastiansen F, Eliassen S, Ginot V, Giske J, Goss-Custard J, Grand T, Heinz SK, Huse G, Huth A, Jepsen JU, Jørgensen C, Mooij WM, Müller B, Pe'er G, Piou C, Railsback SF, Robbins AM, Robbins MM, Rossmanith E, Rüger N, Strand E, Souissi S, Stillman RA, Vabø R, Visser U, Deangelis DL (2006) A standard protocol for describing individual-based and agent-based models. Ecol Modell 198(1–2):115–126

Grimm V, Berger U, Deangelis DL, Polhill JG, Giske J, Railsback SF (2010) The ODD protocol: a review and first update. Ecol Modell 221(23):2760–2768

Jongdee B, Pantuwan G, Fukai S, Fischer K (2006) Improving drought tolerance in rainfed lowland rice: an example from Thailand. Agric Water Manage 80:225–240

Lacombe G, Naivinit W (2005) Modeling a biophysical environment to better understand the decision-making rules for water the use in the rainfed lowland rice ecosystem. In: Trébuil G, Bousquet F, Hardy B (eds) Companion modeling and multi-agent system for Integrated natural resource management in Asia. Los Baños,IRRI, pp 191–210

Naivinit W (2009) Companion modelling to analyze the land/water use and labour migration interactions in Lam Dome Yai Watershed, Lower Northeast Thailand. Cotutelle thèse de doctorat en géographie humaine, économique et régionale de l'Université Paris Ouest Nanterre-La Défense and PhD in Agricultural Technology, faculty of Science, Chulalongkorn University, Thailand. http://cormas.cirad.fr/pdf/Naivinit_2009_PhD.pdf

Naivinit W, Le Page C, Trébuil G, Gajaseni N (2010) Participatory agent-based modeling and simulation of rice production and labor migrations in Northeast Thailand. Environ Model Softw 25:1345–1358

Smajgl A, Brown DG, Valbuena D, Huigen MGA (2011) Empirical characterisation of agent behaviours in socio-ecological systems. Environ Modell Softw 26(7):837–844

Trébuil G, Dufumier M (1993) Regional agrarian systems and sustainability of agricultural production systems in Thailand. J Asian Farming Syst Assoc 1(4):557–568

Trébuil G, Harnpichitvitaya D, Tuong TP, Pantuwan G, Wade LJ, Wonprasaid S (1998) Improved water conservation and nutrient-use efficiency via subsoil compaction and mineral fertilization. In: Ladha JK, Wade LJ, Dobermann A, Reichardt W, Kirk GJD, Piggin C (eds) Rainfed lowland rice: advances in nutrient management research. Los Baños, IRRI, pp 245–256

Voinov A, Bousquet F (2010) Modelling with stakeholders. Environ Modell Softw 25(11):1168–1281

Walker B, Carpenter S, Anderies J, Abel N, Cumming G, Janssen M, Lebel L, Norberg J, Peterson GD, Pritchard R (2002) Resilience management in social-ecological systems: a working hypothesis for a participatory approach. Conserv Ecol 6(1):14. http://www.consecol.org/vol6/iss1/art14

Chapter 12
Building Empirical Multiagent Models from First Principles When Fieldwork Is Difficult or Impossible

Armando Geller

12.1 Introduction

This Chapter informs the reader about how to create and parameterize empirical multiagent models from first principles when fieldwork is difficult or impossible to conduct and data is primarily of qualitative nature. Empirical multiagent models have become ever more popular over the last decade (see Geller et al. 2011a, b; Gurung et al. 2006; Janssen and Ostrom 2008; Latek et al. 2011; Polhill et al. 2010, to name a few recent publications). While informing models using statistical and geospatial data can orient itself on more established techniques and standards (see for example Berger and Schreinemachers 2005; Mussavi Rizi et al. 2012), methodological challenges persist in regards to using qualitative data for informing and parameterizing models. Protocols such as ODD (Grimm et al. 2006) are welcome and helpful devices—and hence used in this Chapter—but qualitative data comes with its own peculiarities. The most important of which is, for modeling purposes, that qualitative data tends to inform the logic of agent behavior. The emphasis I thus put on qualitative data to make model design decisions based on evidence and first principles will be reflected by soft adaptations of the ODD protocol. Arguably this may amount to a deeper insight the Chapter is providing: Whereas the usage of such frameworks as ODD increases model reliability, validity is built using qualitative empirical data for informing and parameterizing the agent and model behavior.

This Chapter presents and discusses the case of modeling power in Afghanistan where primary semistructured interview data and secondary narrative sources were

The notion of first principle and the argument of this paper are predicated on the conviction that empirical modeling should start with observation and description of the case, including its agents, to be studied (cf., Edmonds and Moss 2004; Moss 2002; Moss and Edmonds 2005).

A. Geller (✉)
Scensei, LLC and George Mason University, Fairfax, VA, USA
e-mail: armando@scensei.com

A. Smajgl, O. Barreteau (eds.), *Empirical Agent-Based Modelling – Challenges and Solutions*, DOI 10.1007/978-1-4614-6134-0_12,

used to inform and parameterize the model. The Chapter is intended to provide the reader with enough procedural information to replicate the model information and parameterization process.

The core problem in any empirical multiagent modeling is to align the model ontology with the (implicit or explicit) data model. Ultimately, the data should be able to speak to agent behavior (which includes cognition) and agent environment. Only rarely is data collected with the intent to inform multiagent models, and multiagent modelers regularly need to work with whatever data is available to them. Consequently, selecting adequate data is pivotal; and building an argument for why (in- and output) data speaks to a given model ontology remains a fastidious task. Once the argument is made convincingly, data needs to be translated into the model. The grammar of model building (and indeed science) requests that all this happens intelligibly and reproducibly.

A wide range of multiagent architectures exists from which modelers of social phenomena can choose. The extent to which agents are placed in real world contexts and models are informed by real world data is entirely up to the researcher; but increased empirical expressiveness and richness, one would assume, go hand in hand with increased behavioral and contextual realism. What agents see, hear and feel; how they reason over the perceived information given purpose, incentives and constraints; and what actions they select based on their reasoning should be empirically specified. Depending on educational background and research design scientists may be more interested in informing and parameterizing empirically *what* agents reason about or *how* they reason about it, or both. In what follows the focus will lie more on what agents reason about and how they act upon it without, however neglecting the importance of cognitive processes.[1]

Even without applying especially devised data collection protocols is qualitative data often sufficiently differentiated to inform model ontology and sufficiently detailed to inform and parameterize agent architecture, as was for example shown in work by Alam et al. (2010) and Geller et al. (2011b). In fact, the data is often so rich in behavioral content—usually less so in numeric content—that traditional parameter and behavior sweeps are not necessary because the data is unambiguous.[2] Sweeping through the behavior space would mean to change behaviors because only unambiguous data is available.[3] Both, making ad hoc assumptions based on ambiguous data and fiddling with unequivocal data without good reason is no good practice.[4]

[1] For those interested in an extensive discussion of reasoning in multiagent modeling I refer to Latek 2011 and Sun 2008.

[2] Sweeping instead is performed to account for varying data sources, creating multiple bodies of knowledge instantiated as models and simulations.

[3] In the Anasazi model (Kohler et al. 2005), for example, behavior space was varied exactly because the researchers wanted to create plausible behavioral and environmental hypotheses and explanations for why this Puebloan people society suddenly collapsed. And in policy oriented work behaviors are varied to generate and explore what-if scenarios to inform policy makers (Lempert et al. 2006).

[4] Geller et al. (2011b) show that the same also applies to statistical data. Why sweeping the parameter space when the data is known?

12.2 Model Description Using the ODD Protocol

Let me first provide a general overview of the *purpose*, *state variables and scales*, and *process overview and scheduling* of the so-called *qawm* model. Everyday life in Afghanistan is a lot less regulated by government institutions than in the industrialized world. Informal power as an applied concept to organize individual and social behavior plays a pervasive role (Azoy 2003). Yet, many aspects of it remain elusive. Scott Moss and I wanted to capture these informal power relationships and the reasoning on the agent level that bring them about as part of a computational model (Geller and Moss 2008).[5] The reason we referred to this work as the *qawm* model is because the Arabic word *qawm* is used in Afghanistan to denote (opportunistic) solidarity networks.

In particular we were interested in the formation of power structures among Afghan stakeholders for whom *qawm* are a means of acquiring, maintaining and increasing power. The modern—that is mutated by 30 years of conflict and international aid—functional rationale of *qawm* can be understood in the broader notion of neopatrimonialism. Mousavi (1997) refers to *qawm* as complex interpersonal networks of political, social, economic, military and cultural relations. Afghan social structure does not take the form of a unified hierarchy, nor does an individual *qawm*. However, each *qawm* has a *primus inter pares* who competes with other *primi inter pares* as well as with *qawm* internal rivals for manifold reasons (Azoy 2003). The social interactions within *qawm* determine a pattern of actions that could be described as episodic clusters of cohesion building and dissolution. Such *qawm* level behavior leads to interactions among the *qawm* that cause episodic alliance building or conflict of unpredictable magnitude, duration and outcome. It is these *qawm* that we can refer to as the higher-level entity in the model. There are no other higher-level entities represented for our agents are only embedded in an abstract geospatial setting and power is represented mainly as a socio-mental reification.

The low-level entities of our model are agents representing individuals, the state of which can be characterized using the following variables: role, ethnicity, kinship, religion, political orientation, and *hisiyat* and *e'tibar*, roughly translating from Dari into the English words character and credit respectively. Section. 12.3 will provide more details on these state variables and in particular on *hisiyat* and *e'tibar*.

Agents in the *qawm* model have no purpose due to a daily pulse of life. They do not have to eat and work. It is instead assumed that they are inherently social and have to cooperate to master a task. In these interactions agents accumulate and redistribute social and material resources—processes depending on their identity, how much resources they have themselves and how they are embedded in their social network relative to others. Consequentially, there is no bio-social notion of time. In every time step all agents are activated in random order. The scheduling, however, is determined by the declarative framework, which we chose to implement the simulation in: The sequence of rules that will fire and the particular instantiations

[5] The following sections describing the model are adapted from Geller and Moss 2008.

of their variable values are determined only as the program is running for rules fire only when certain conditions are met. Again, Sect. 12.3 will give more detail on the declarative framework.

The next step in the ODD protocol is the description of *design concepts*. Two design concepts stand in the foreground: a declarative framework and so called *endorsements*. Allow me to describe these so important concepts for the understanding of our work—they strongly affect how agents are informed, as will become clear in Sect. 3—, before returning to the classic ODD protocol.

Declarative programs are written in a way that express what should be accomplished (in our case describe conditions on a solution to a social organization problem), but not how it should be done. That is left to the compiler, which in our case was JESS. Instead of us untying the behavioral consequences of the rich narrative form in which most of our data was, we had JESS deal with it for us. Since order ("preference") and replication ("weight") should not play a role in declarative programming more or less importance was not attributed to any rule as a result of programing.

Our understanding for the *qawm* model was that a program is declarative if there are a set of statements on a database, rules have a set of conditions which are statements with some values left open as variables, and consequents exist which are another set of statements. When all of the statements in the conditions of a rule are matched by statements on the database, then the variables are given their specific values from the database statements and the consequent statements are added to the database. When a set of conditions are satisfied and a rule fires (i.e., puts its consequents on the database), then the state of the environment as represented by the database is changed and perhaps other rules will now be able to fire and so on until all rules have fired and no further matches of conditions can be found on the database. The sequence of rules that will fire and the particular instantiations of their variable values are determined only as the program is running. The sequence of actions represents the process of agent behavior and leads in each case to a new state of the environment. If all agents are implemented declaratively, then they will be changing the state of the environment for one another and the pattern of rules, and therefore actions of all the agents taken together will be influenced by one another.

The core of our agents is cognitive in nature and tells us something about what an idealtypical "Afghan agent" reasons about when thinking about power. In the *qawm* model we used a technique called *endorsements* to implement in computer code our qualitative knowledge of these reasoning processes (see for the following in more detail Alam et al. 2010). Endorsements capture the reasoning process of one agent, the endorser, about another agent, the endorsed. Endorsements can be considered as labels, which agents use to describe certain aspects of other agents in a subjective manner. These labels can be affirmative and even positive like is-kin, is-neighbor, is-friend, similar, reliable, and capable; or non-affirmative and even negative like non-kin, unreliable, incapable and untrustworthy. Some endorsements are static in that, once identified, they do not change over the course of the simulation (e.g., is-kin), while others are dynamic and may be revoked, replaced, or dynamically adapted according to an agent's experiences.

To assess endorsements, agents rely on a so-called endorsement scheme, which associates each label with a weight to express how much value an agent assigns to this particular aspect related with another agent. Weights are modeled on an ordinal scale as integer numbers between 1 and n (n being a real number) for positive labels and −1 and $-n$ for negative labels, respectively. This allows for computing an overall endorsement value E for an agent as depicted in Eq. 12.1 where b is the number base and w_i is the weight of the i_{th} endorsement token.

A social scientific interpretation is that the base b represents an agent's general disposition. This can be an indicator for an agent's nature, e.g. extreme or moderate; it is also an indicator for the agent type or role, such as academic, businessperson, sport star, etc. The weights w, on the other hand, define the importance of the context, therefore assigning more or less weight to the base depending on a particular situation. Yet, there is no social meaning that pertains to the absolute values. Rather, values indicate a tendency relative to other values, hence why they are of ordinal character. In the case at hand, these values have not been empirically derived, although they could, at least in regards to the tendency they give expression to.

From a processual perspective, the endorser's endorsement scheme is projected onto the endorsee. If an agent A_1 wants to evaluate whether an agent A_2 should be endorsed or not, A_1 has no objective base to rate A_2 and its labels respectively and to make a decision based on this information. What A_1's individual endorsement scheme tells A_1, however, is how important some or all of A_2's labels are for A_1. If this is done for each of the endorsed agent's labels, E for the endorsee, in the given example A_2, can be calculated according to Eq. 12.1.

$$E = \sum_{w_j \geq 0} b^{w_i} - \sum_{w_j < 0} b^{|w_i|} \tag{12.1}$$

E allows the endorser to choose the preferred one(s) among a number of agents it endorses. The process of choosing an agent is embedded in an agent's context, i.e. the agents visible or known to it. Relying on endorsements allows an agent to find the agent most appropriate to it within its context. This implies that the chosen agent may not be preferable to differently embedded agents with a different endorsement scheme.

Back to classic ODD: What is emerging from individual level interactions are social relationships and social structure, in other words *qawm*. But agents do not adapt nor are they fitness-seekers or anticipatory.[6] However, agents are equipped with a fairly rich cognitive architecture in terms of social dimensions they reason about; the state variables mentioned afore. Agents deliberately take them into account in their decision making via *endorsements*. The interactions agents engage in are thus assumed to be socio-cultural relationships: trust, friendship, consanguinity, etc., and agents form collectives, that is *qawm* on the basis of these relationships. Most of the data collected from the model is data about these social networks. And

[6] Alam et al. (2010) describe how to make endorsements an adaptive concept through introducing memory.

while stochasticity plays a role in the order in which agents are activated and values are assigned to state variables at agent level during initialization, the declarative nature of the approach requires no explicit randomization.

The model has, due to its conceptual simplicity, relatively little *details* to report about. Agents are instantiated using either random values to represent cognitive traits (*endorsements*) or empirical values and informed estimates for population numbers and demographics (roles, religion, ethnicities). The standard parameter values are: The simulation is spatially based on a 50 × 50-cell, 2D-grid topology and each cell can be inhabited by one agent only. The total number of cells is assigned to four ethnic regions based on ethnic ratios. The total number of agents ranges around 200, with a usual distribution chosen of 6 politicians, 6 religious leaders, 6 businessmen, 6 organized criminals, 6 commanders, 10 drug dealers, 35 drug farmers, 35 farmers, 70 civilians and 28 warriors. Each agent is randomly assigned a number of kinspersons, a politico-military background and a Moore neighborhood. It is furthermore assumed that elites have a vision (the number of cells they can "see") of 11 × 11 cells; drug dealers have a vision of 9 × 9 fields; and ordinary agents have a vision of 5 × 5 cells. The range in which agents can move around the grid is proportional to the size of their vision.

In general, percentages accepted for the main four ethnicities are 40 % Pashtun, 25 % Tajik and Uzbek each, and 10 % Hazara. Because Pashtun, Tajik and Uzbek are all Sunni we are left with a Sunni to Shia ratio of 9:1.

There is an exogenous resource that is lognormally distributed among agents and that can be accumulated and redistributed.

The majority of the land is assumed to be rural. There are three cities. Rural areas are rather homogeneous in terms of ethnicity and religion, whereas cities are a multicultural space. Some agents belong only to rural spaces, such as farmers and drug farmers; some only to the city, such as organized criminals and businessmen.

If I would have to categorize the *qawm* model against the background of the decision tree described in Chap. 1, then it would be a small *N* model in which the entire population (or a reasonable sample thereof) cannot be studied using fieldwork due to security restrictions and availability of stakeholders.[7] Instead a combination of fieldwork, expert knowledge and extant secondary data was used to parameterize agent type, behaviors and attributes. Conceptually that puts us somewhere in between of cases 14 and 16 in the parameterization sequence.[8] However, because we chose to apply a declarative modeling technique, attribute and behavioral data elicitation methods are kept at a minimum in order to preserve the character of the raw data. This is further reflected by the chosen cognitive architecture, *endorsements*.

---

[7] That experimental work can however be conducted even in very difficult security conditions is shown by Geller et al. 2012.

[8] And if I would have to pin it down to a particular case, then 15 would probably fit best.

12.3 Parameterization Details

Speaking in the language of the Characterizing and Parameterizing (CAP) framework (see Chap. 1 in this volume; Smajgl et al. 2011), the parameterization sequence applied by us is as follows: The identification of the different agents (M1) was conducted using expert knowledge and extant (in the sense of extant ethnographic data) participatory observation; (qualitative) agent attributes (M2) and behavioral responses (M3) were obtained using interviews, expert knowledge and extant participant observation; where agent types were not already clear in step M1, they were developed from attributes (M4a) and behavioral components (M4b) using expert knowledge and extant participant observation; assignment of agents to agent types on population level (M5) was conducted using no particular scaling technique, but was driven by expert knowledge and computational design.

Yet, the CAP framework is not easily reconcilable with the declarative framework applied by us. The purpose of the model, to capture informal power relationships and the agent reasoning that brought them about in a social simulation, and the nature of the data we held in our hand made us conclude that an evidence driven and declarative modeling approach would be most receptive to our intentions. Because such an approach intentionally keeps us close to the raw data no formal elicitation and parameterization procedure was followed.[9] Together with the endorsement cognitive architecture it established a first principle based framework for empirically informing the agent types and their attributes and behaviors.

Using declarative modeling and endorsements, parameterization of the simulation is performed against the behavioral and structural logic—expressed as agent reasoning, networks of relationships and storylines—of the "case" as indicated by the data rather than specific handpicked values. In order to retrace parameterization one needs to understand that logic and the data that stands behind it.

Plenty of extant secondary data provide descriptive accounts of what *qawm* are, but only rarely are they accompanied by behavioral data of the sort needed for informing an agent. Azoy (2003) is an exception, providing a wealth of behavioral information in narrative form (participant observation). In combination with our own expert knowledge we were able to develop empirically informed idealtypical profiles of the agents and the society they are living in that was about to inform our computational model. The agents and the relationships between the agents are represented in Fig. 12.1 (step M1 in CAP). There are ten actor types: politicians, religious leaders, commanders ("meritocratic title" for a militia leader), businessmen, warriors, civilians, farmers, drug farmers, organized criminals and drug dealers. Powerful agents, that is politicians, religious leaders, commanders, businessmen and organized criminals, form affiliations with each other, while "ordinary agents," that is civilians, warriors, farmers and drug farmers, form patron–client relationships with strongmen. In the model each actor has its distinct role, whereas in reality

[9] Parts of the modeling philosophy behind our approach, KIDS (Keep It Descriptive Stupid), are described in Edmonds and Moss (2004). It is important to mention our modeling philosophy here, because it defines in parts how we approach parameterization.

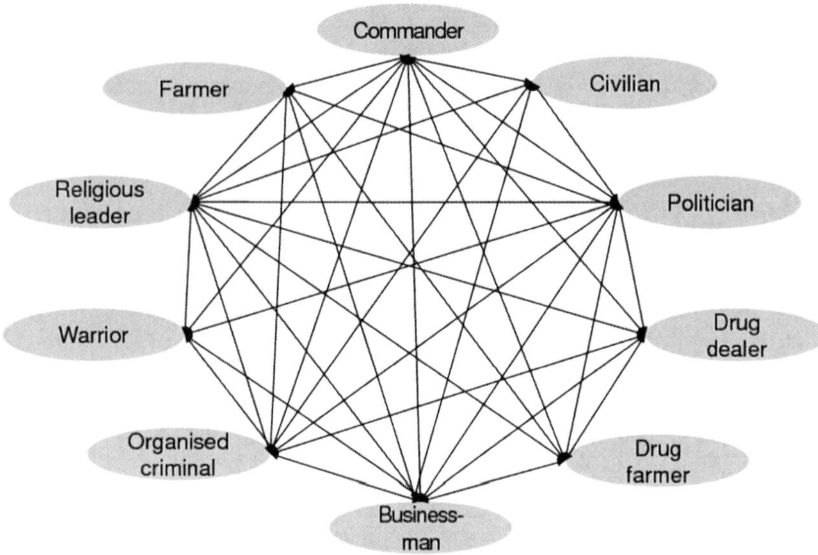

Figure 12.1 Idealtypical representation of *qawm*

actors may incorporate a variety of roles. These choices are based on our own understanding of the case as developed over time in our own fieldwork and reading the relevant literature; and as a result of direct feedback from interviewees who were invited to comment on Fig. 12.1.[10]

Our own data and secondary data such as Azoy (2003) emphasize the importance of ownership as a source of power. Traditionally, ownership can be defined as land, access to water, livestock and women. References to landownership were made often during interviews, whereas water and women have never been mentioned. Ownership of livestock was mainly mentioned to serve reputational means. The interview data and observations made during our field trips suggest that a modern comprehension of ownership has become more materialistic and less subsistence oriented; mundane symbols of power such as money, houses and cars have increased in importance. Thus, we decided to endow our agents with some not further specified material resource. Because agents are created as being powerful or ordinary by definition material resources do not have to be distributed explicitly; agents are created possessing them, or not. We did not intend to create a synthetic representation of the real Afghan population or a sample thereof and using a rough, but plausible approximation for the distribution of wealth was deemed sufficient by us. Based on accounts in the literature (UNODC 2006; Wily 2004) and our own understanding of

[10] None of our interviewees disagreed in principle with the representation in Fig. 12.1 in terms of which agent types we chose and how we characterized them as being either powerful or ordinary. We therefore did not deem it necessary to return to the idealtypical representation of *qawm* and make changes to it.

the case we chose this distribution to be "fat tailed", meaning that a large number of agents are poor, while a small number of agents is fairly rich.

The model reveals its real richness in terms of agent behaviors and attributes (steps M2 and M3). To get at these we attempted to place ourselves into the agent mind in order to better understand what an idealtypical "Afghan agent" could be thinking about when reasoning about power. Existing ethnographic work such as Azoy (2003) was useful, but the bulk of the data I collected myself. I interviewed 34 Afghans between 2006 and 2008 that can be characterized as urban elites. 30 of them were male, 4 female. Among them were Pashtun, Tajik, Hazara, Uzbek and Aimaq from Kabul, Kandahar, Mazar-e Sharif, Herat, Bamyan and other cities. Some of them were Mujahedin in the Jihad against the Soviets, some of them sided with the then Communist Government, and some of them later became Taleban. They belonged to political, military, economic, religious and intellectual elites. I considered them elites because at the time all of them were in a position to influence the situation on the national or regional level. This is a sample that is by far more characterized by opportunity than representativeness, but the 1–2 h long semistructured interviews gave me a good understanding about how Afghan urban elites reason about power. I structured the interviews around the following three questions:

- Under what conditions would you label someone as powerful?
- Having labeled someone as being powerful, how would you expect that person to behave?
- Having labeled someone as powerful, how would you expect yourself/others to behave towards that person?

The following examples should give an idea of how a sanitized version of the interview data would look like: An interviewee would tell me that if a politician is in need of military protection, he approaches a commander. In return, a commander receives political appreciation by mere cooperation with a politician. Or an interviewee would tell me that if a businessman wants to be awarded a governmental construction contract, he relies on a politician's connections. In return, the politician receives a kickback. Or an interviewee would tell me that if a politician wants beneficial publicity, he asks a religious leader for support. The religious leader, in return, becomes perceived as a religious authority. Or an interviewee would tell me that if a warrior seeks subsistence for his family, he lends his services to a commander, who in return provides him with weapons, clothes, food and/or money. Or an interviewee would tell me that if an organized criminal wants to carry drugs, he relies on a businessman's transport business, and the businessman in return receives a share of the profit from the sold drugs. Or an interviewee would tell me that if a drug farmer needs protection for his poppy fields, he affiliates himself with a commander, who in return receives a tithe of the profit from the drugs sold to a drug dealer.

We then overlaid the idealtypical *qawm* in Fig. 12.1 with narratives like these. But unlike more established behavior elicitation and typology development approaches where some kind of technique is applied to condense rich data into an idealtypical form we chose a declarative approach where the behavioral rules are directly coded from the raw interview data in order to preserve the original behavioral

Table 12.1 Behavioral rules for the agent type *commander*

No.	Rule
1	Default-daily-payment-commander-to-warrior
2	Affordable-warriors
3	Commander-recognises-newly-recommended-warrior
4	Commander-recognises-known-recommended-warrior
5	Commander-collect-warriors
6	Commander-endorse-warrior-as-reliable
7	Commander-endorse-warrior-as-unreliable
8	Commander-endorses-warrior
9	Commander-endorses-businessmen
10	Commander-sends-message-to-best-endorsed-businessman
11	Commander-endorses-politicians
12	Commander-sends-message-to-best-endorsed-politician
13	Commander-asserts-trustworthiness-affiliation-with-politician
14	commander-sends-message-to-answer-politician-protection-request
15	Commander-endorses-religious-leaders
16	Commander-sends-message-to-best-endorsed-religious-leader
17	Commander-sends-message-to-answer-religious-leader-spiritual-leader-request
18	Commander-sends-message-to-answer-businessman-protection-request
19	Commander-sends-message-to-accept-businessman-protection-request
20	Commander-selects-warriors-to-approach
21	Commander's-warrior-endorsement-values
22	Commander-offers-to-hire-warrior
23	Commander-invests-money

logic. M4a, b and M5 in CAP also exist in our procedure, but have a function that is less synthesizing and more inclusive. The coding process also triggered requests for further or more specific information that would allow us to complete the reasoning logic. We ended up with 173 behavioral rules. A rule sample for the agent type *commander* is presented in Table 12.1. The rules have self-explanatory names.

So far the data and rules presented tell us something about relational and situational instances, but little about the categories on the basis of which decision are made is said. To inform these mental categories we used a socio-cultural concept taken from Azoy (2003) and enriched and completed it where necessary with our own data.[11] The concept consists of the two notions *hisiyat* and *e'tibar*, roughly translating into character and credit respectively. *hisiyat* denotes qualities such as piety and wisdom; *e'tibar* is about meritocracy. Someone powerful must dispose of both, *hisiyat* and *e'tibar*. *hisiyat* and *e'tibar* have to be further differentiated should they become tangible attributes about which an agent can reason in meaningful ways. The agent attributes that we find in our own data for *hisiyat* and *e'tibar* are listed in

[11] Nothing changed in principle to our approach as described above with regard to M2, M3, M4a, b and M5.

Table 12.2 Reasoning categories of an "Afghan agent"

	Static	Dynamic
hisiyat	Intellectual/non-scholarly	Loyal/disloyal
	Shared-ethnicity/different-ethnicity	Trustworthy/untrustworthy
	Shared-religion/different-religion	Is-neighbor/non-neighbor
	Is-kin/non-kin	Pious/sinful
	Politico-military-background	
e'tibar	–	Reliable/unreliable
		Successful/unsuccessful
		Capable/incapable

Table 12.2. Some of these attributes are dynamic, that is they can change over the "lifetime" of an agent; others are static and remain the same. In other words, some parts of an agent's identity are given, such as who their kin are and what religion they have (they cannot convert). Other parts of their identity are evolving, such as their trustworthiness or their success. This means that some categories of an agent's identity can be informed empirically, using for example estimates of ethnicity and religion probabilities in the Afghan population (see Sect. 12.2). Other concepts are more difficult to inform empirically in a meaningful way other than that they exist mentally and play a role when reasoning about power, such as piety, and trustworthiness. As explained above, the model's goal was not to produce a quantification of power in Afghanistan (which I doubt that it can be done), but to create a more formalized notion of how Afghans reason about power.

While these categories are a mental concept for the endorser, they are part of the identity, the Self of the endorsee. This concept of Self is built around the endorsement scheme, because it is what agents reason about, i.e. about the identity of other agents. Agent identity is based on some demographic data, but mainly informed by what we know from our own data and the literature about what agents reason about when they reason about power. While values are attributed to reasoning categories, they are strictly of ordinal type and not more should be interpreted into them than mentioned above. Sweeping the parameter space in this case may have a technical value for the simulation, but only little socio-scientific meaning.

Instead, Scott Moss and I constantly monitored the agent ruleset that serves as model input and the model output in the form of storylines and compared them against our knowledge and understanding of the case. In fact, we deliberately devised our interactions as roleplay between subject matter expert and modeler. In addition to this we presented to the interviewed Afghan elites the agent ruleset and the storylines produced as output by the simulation and invited them for comments. This participatory approach to "parameterization" helped us better aligning the model and the simulation with the evidence in our hand.

12.4 Experiences and Lessons

The *qawm* model as presented forms part of a class of models that has been innovated in the Centre for Policy Modelling at Manchester Metropolitan University. In the development of these models priority has always been given to the empirical evidence (Moss 2002). The experiences and lessons I draw from the effort have to be understood against this background.

Modeling power in Afghanistan using a combination of participatory, declarative and rich cognitive modeling generated, I believe, a more evidence-based and "thicker" understanding of the case at hand. Data of ethnographic quality were mixed with subject matter knowledge to parameterize the model in terms of agent types, behaviors and attributes. Such a procedure goes beyond a classical understanding of parameterization, increasing construct validity significantly. A valuable side effect of such a research design is the facilitation of dialogue between participants and thereby mediation of concepts.

Declarative representations of agents have a number of virtues in terms of ease of development as new evidence becomes available and in terms of yielding comprehensible outputs stored as statements on databases. The declarative approach allowed us to translate the evidence at hand into computer code without giving up its richness and expressiveness because in declarative programing the order of statements and expressions, and the replications of statements should not affect program semantics. Furthermore, each time a rule fires the conditions for why the rule fired and the new state of the environment as represented by the database can be checked against the evidence.

The strength of the applied parameterization approach undoubtedly is its expressiveness. Yet, this comes at a cost. The pragmatic nature of the parameterization sequence has to rely on the domain knowledge of the modelers. Lengthy arguments over what needs to be included and what not due to a lack of formalization and extensive use of qualitative data are inevitable.[12] And despite its expressiveness the model remains fairly simple (somewhere between 0 and 1 on Axtell and Epstein's (1994) scale) with, however, a quite large number of degrees of freedom. In other words the modeling team's hands will get dirty and the team necessarily needs to have a unified philosophy of modeling and be willing to work disciplined.

A note is also needed on arbitrary, that is non-empirical parameter values. It is a well-known fact that results derived from simulations are strongly dependent on the chosen parameter values.[13] Yet little information of socio-scientific relevance is added by the technical exercise of sweeping arbitrary parameter spaces and no meaningful empirical interpretation can be derived from it. What we find is that our results are stable across a relatively broad parameter space, meaning that our findings are not driven by a distinctively chosen set of parameters.

[12] Indeed, this is one of the main reproaches brought against the KIDS approach.

[13] See for an important, yet often neglected paper Centola et al. 2007.

Declarative modeling in combination with rich qualitative data make parameterization intelligible—everyone interested can inspect the agent rules and trace them back to the qualitative data from where they have been derived from—, but also tedious to replicate.[14] The outcomes for the model as a whole are, in these circumstances, difficult to predict with any exactitude. While the model was well suited for its intended uses within its scope of empirical applicability, due to the specificity inherent to the chosen approach generalizability is limited and probably never goes beyond the purpose for what the model was designed for.

12.5 Conclusions

Parameterization focused simply on point values is unnecessarily narrow and probably outdated. Recent work in fusing qualitative and quantitative data suggest taking a much broader stance that includes taking into account information usually mediated by qualitative data. This kind of data provides insight about the meaning and logic of a particular situation, action, relationship and so forth. Understood as such parameterization becomes an integral part of a model's internal validity.

In the case at hand we looked at power relations between actors. Combining participatory, declarative and rich cognitive modeling approaches worked well for us because it allowed us to integrate a diverse and rich dataset into a computational social simulation. Ultimately, the decision to work with the selected data and the agent types, behaviors an attributes derived from it worked for us and the questions we were interested in. A claim to generality is made only with regard to how we selected the data and used it for modeling purposes, namely through detailed knowledge of the case and stakeholder involvement, not with regard to content, which is context-sensitive. Someone interested in say power in a gender context in Afghanistan may want to add women in various roles to the model.

The modeling work presented shows that data specifically collected for multiagent modeling and simulation purposes, the presentation of these data adequate to their needs in multiagent modeling and a rigorous data translation process for parameterization go hand in hand. Overall, collecting data particularly for multiagent modeling and simulation would simplify data translation because it narrows the gap between data model and multiagent model ontology. Ultimately, developing empirical multiagent models from first principles when fieldwork is not possible or difficult (as in our case) makes hands dirty because hard fieldwork and long arguments are needed instead of elegant mathematics.

[14] Note that a rule-based system like JESS makes replication particularly difficult because the sequence of how the model runs is defined by the compiler and explicitly not in the hands of the programmer.

References

Alam SJ, Geller A, Meyer R, Werth B (2010) Modelling contextualized reasoning in complex societies with "endorsements". J Artif Soc Soc Simul 13(4). http://www.jasss.soc.surrey.ac.uk/13/4/6.html

Axtell RL, Epstein JM (1994) Agent-based modeling: understanding our creations. The Bulletin of the Santa Fe Institute. Winter:28–32

Azoy GW (2003) Buzkashi: game and power in Afghanistan, 2nd ed. Waveland, Long Grove

Berger T, Schreinemachers P (2005) Empirical parameterization of multi-agent models in applied development research. Paper prepared for the workshop "Empirical techniques for testing agent based models". Indiana University, Bloomington, USA, 2–4 June 2005

Centola D, González-Avella JC, Eguíluz VM, San Miguel M (2007) Homophily, cultural drift, and the co-evolution of cultural groups. J Conflict Resolution 51(6):905–929

Edmonds B, Moss S (2004) From Kiss to KIDS—an "antisimplistic" modelling approach. In: Davidsson P, Logan B, Takadama K (eds) Multi agent based simulation. Lecture notes in artificial intelligence 3415. Springer, Berlin, pp 130–144

Geller A, Moss S (2008) Growing Qawm: An Evidence-Driven Declarative Model of Afghan Power Structures, Advances in Complex Systems 11(2):321–335

Geller A, Harrison JF, Revelle M (2011a) Growing social structure: an empirical multiagent excursion into kinship in rural North-West Frontier Province. Struct Dyn 5(1). http://www.escholarship.org/uc/item/4ww6x6gm

Geller A, Latek MM, Mussavi Rizi SM (2011b) How corruption blunts counternarcotic policies in Afghanistan: a multiagent investigation. In: Salerno J, Yang SJ, Nau D, Chai S-K (eds) Social Computing, Behavioral-Cultural Modeling and Prediction 2011 conference. Lecture notes in computer science 6589. Springer, Berlin, pp 121–128.

Geller A, Łatek MM, Mussavi Rizi SM (2012) Using social simulation for aid and development analytics during armed conflict: Experiences from Uruzgan, Proceedings of the Third Brazilian Workshop on Social Simulation

Grimm V et al (2006) A standard protocol for describing individual-based and agent-based models. Ecol Modell 198(1–2):115–126

Gurung TR, Bousquet F, Trébuil G (2006) Companion modeling, conflict resolution, and institution building: sharing irrigation water in the Lingmuteychu Watershed, Bhutan. Ecol Soc 11(2). http://www.ecologyandsociety.org/vol11/iss2/art36/

Janssen M, Ostrom E (2008) Empirically based, agent-based models. Ecol Soc 11(2). http://ecologyandsociety.org/vol11/iss2/art37/

Kohler TA, Gumerman GJ, Reynolds RG (2005) Simulating ancient societies. Scientific American July:77–84

Latek MM, Mussavi Rizi SM, Alsheddi TA (2011) Optimal blends of history and intelligence for robust antiterrorism policy. J Homel Secur Emerg Manag 8(1)

Latek MM (2011) On bounded rationality in multi-agent environments. Doctoral Dissertation. George Mason University

Lempert RJ, Groves DG, Popper SW, Bankes SC (2006) A general, analytic method for generating robust strategies and narrative scenarios. Manage Sci 52(4):514–528

Moss S (2002) Policy analysis from first principles. Proc Nat Acad Sci U S A 99(3):7267–7274

Moss S, Edmonds B (2005) Towards good social science. J Artif Soc Soc Simul 8(4). http://www.jasss.soc.surrey.ac.uk/8/4/13.html

Mousavi SA (1997) The Hazaras of Afghanistan. An historical, cultural, economic and political study. St. Martin's, New York

Mussavi Rizi SM, Łatek MM, Geller A (2012) Fusing remote sensing with sparse demographic data for synthetic population generation: an algorithm and application to rural Afghanistan, International Journal of Geographical Information Science 27(5) 986–1004.

Polhill G, Sutherland LA, Gotts NM (2010) Using qualitative evidence to enhance an agent-based modelling system for studying land use change. J Artif Soc Soc Simul 13(2). http://www.jasss.soc.surrey.ac.uk/13/2/10.html

Smajgl A, Brown DG, Valbuena D, Huigen MGA (2011) Empirical characterisation of agent behaviours in socio-ecological systems. Environ Modell Softw 26(7):837–844

Sun R (2008) Prolegomena to integrating cognitive modeling and social simulation. In: Sun R (ed) Cognition and multi-agent interaction. Cambridge University Press, Cambridge, pp 3–26

UNODC (2006) Afghanistan's drug industry. Structure, functioning, dynamics, and implications for counter-narcotics policy. United Nations Office on Drugs and Crime, New York; The World Bank, Washington D.C.

Wily LA (2004) Rural land relations in conflict: a way forward. Afghan Research and Evaluation Unit, Kabul

Chapter 13
Designing Empirical Agent-Based Models: An Issue of Matching Data, Technical Requirements and Stakeholders Expectations

Olivier Barreteau and Alex Smajgl

13.1 Introduction

Despite their diversity, the 11 examples of empirical agent-based model design described in this Volume enable not only a consolidation of the CAP framework described in Chap. 1, but also an exchange of experiences in designing empirical agent-based models. The detailed descriptions of the example modelling processes showcase the methodological diversity and the state of art practiced within the emerging community of empirical agent-based modelling. All these examples have their own limitations as a matter of empiricism that the framework aims to structure. In this final Chapter we discuss effectiveness and robustness of the Characterisation and Parameterisation (CAP) framework, which we revised during the process of editing this Volume. Then, we discuss how the distinction of particular cases performed, which is followed by a discussion on the diversity of methods. Finally, we use the cases presented here (admittedly small in number) to provide some initial insights for the selection of suitable methods.

13.2 Framework Robustness

This book is a new step in the history of the framework described in Chap. 1. This framework was initially designed on the basis of a few examples interpreted by Smajgl et al. (2011), largely derived from their own experience in designing empirical agent-based models. We improved this first version on the basis of a larger sample

O. Barreteau (✉)
IRSTEA, Montpellier, Montpellier, France
e-mail: olivier.barreteau@irstea.fr

A. Smajgl
CSIRO Ecosystem Sciences, Townsville, Australia
e-mail: Alex.Smajgl@csiro.au

A. Smajgl, O. Barreteau (eds.), *Empirical Agent-Based Modelling – Challenges and Solutions*, DOI 10.1007/978-1-4614-6134-0_13,

of examples for empirical agent-based models and feedback received in response to the earlier version.

This book put the framework to a test by asking the key modellers of 11 empirical agent-based models to apply the framework. Among them, ten had not been involved at all in the design of the initial or the revised framework.

Three criteria are used to discuss the effectiveness of the framework. First, a framework, by definition needs to be generic and, hence be applicable across a diverse set of case studies. Second, the framework needs to effectively structure the process. Third, the framework needs to force modellers to document all details necessary to allow for comparisons.

The selection of examples presented in this Volume constitutes a very diverse set of empirical modelling processes. The authors, who come from diverse communities, were able to apply the CAP framework suggesting that the framework is sufficiently generic. We believe that the framework performed well because it allowed descriptions of the CAP process for a number of very diverse approaches; ranging from cases with an empirical basis rooted in existing data to cases that employ *ad hoc* experiments, or targeted field work, aimed at collect required data. The CAP framework also performs well across the considerable discrepancy in the size of models and populations investigated. While most large N examples require statistical processing at some point, cases of small N can afford to pay more attention to contextual detail and individual peculiarities. The CAP framework suits both ends of the spectrum, large and small N situations. Additionally, the CAP framework can be applied independent from how many persons or households of the target population the modeller aims to simulate.

Regarding the framework structure, our second criterion, we conclude that the framework captures all relevant steps in the characterisation and parameterisation process. All examples could be described as a sequence of methods meant to:

- characterize the model
- specify the attributes
- specify the behaviors
- specify the types
- generate a virtual population.

We conclude that the framework performed well because it did not disregard a practised step of the characterization and parameterisation process, suggesting it is sufficiently generic. Additionally, the principle sequence of steps seems robust although some examples, such as Chaps. 3, 6, 10, and 11 conducted at least parts of the process in multiple iterations. This emphasises the need to understand the framework not as a strict sequence of steps but as a principle structure that can require iterative refinement. These iterations can also be performed within the CAP box (see Fig. 1.1 of Chap. 1) and not only as part of the model assessment feedback loop. All iterations aim to improve model robustness by critiquing model assumptions as well as providing information on:

- further sources of relevant data;

- structural modification of model characterization;
- change in up-scaling techniques to validate the virtual population generated.

The third criterion is potentially the most critical one, as the framework should force modellers to document all technical details to a level that allows the reader to compare different modelling processes and also to replicate the process. We conclude that the CAP framework provides an effective structure for describing all necessary details in a structured and comparable manner. Based on the structured description empirical agent-based models can be compared. Ultimately, such a comparative evaluation allows identification of CAP methods that contribute most effectively to the robustness of an empirical agent-based model. However, as the effectiveness of CAP methods depends on the modelling situation this comparison can only be applied within similar situations, which we distinguish into 16 particular cases.

The structured details provided in the description of the model design process make these examples easier to replicate. Strict replication is not relevant for empirical ABM, since the whole case is modified when performing and publicising simulation outcomes, changing the conditions for replication. The replication of tested CAP approaches is mainly useful for a new model development, as an example for a comparable modelling situation can provide guidance based on the previous experience. The classification of cases distinguishing modelling situations as proposed in Chap. 1 aims to guide the reader; finding an example for a similar modelling situation provides the newcomer with a comprehensively described example of a CAP approach and the lessons learnt from it. This enables the reader to implement the same approach or at least take this example as a starting point for investigating what CAP approach could best suit the particular modelling situation. The detailed description can also be the basis for an ex-post assessment of the simulation and modelling process. It enables the modelling team to step through the detailed sequence of the model design process and identify potential framing induced by the design itself.

This Volume covers 8 out of the 16 cases depicted in the decision tree (Fig. 1.2) in Chap. 1. The diversity captured by this decision tree features differences according to the population size, as well as the possibility to connect to data. The further on the right in the decision tree (small N and model based on partial knowledge of the population) the more difficulties were met by the authors in describing their CAP process with the structure proposed in the framework—because of adjustments made in the "model assessment" feedback loop or because of a different way of connecting with data. In Chap. 12, Geller points out that his way of modelling would not need parameterisation since he uses raw data suiting a modelling philosophy of a declarative language, SDML. We believe that this falls under parameterization, but in the qualitative way mentioned in Chap. 1. While traditionally in many modelling techniques parameters are specified quantitatively, agent-based modelling allows state variables (or entities' attributes) to be defined in strings (or verbal descriptors). Nevertheless, such qualitative (or categorical) definition of states still specifies parameters. For example the population of rules available to agents and the categories used to describe an agent's type constitute parameters of the model, as well as their proportions within categories. These parameters are elicited from the considerable amount of empirical knowledge gathered by the author or from previously existing ethnographic data. In empirical agent-based modelling, parameterisation is not only an issue of pinning down numbers.

The outcome of the framework-based description of each case features the same level of diversity as the cases themselves. Besides the differences in using quantitative and qualitative parameters, we found diversity in sequences as well as in methods used for each step. While all cases start with M1, the path afterwards is variable as it depends on the availability of data and its relation to the target system. The last step, M5, is not always described. In some cases with small N (e.g. Le Page et al. in Chap. 11) the process is finalised after M2 and M3 are completed as the whole virtual population is already specified with no further need to upscale.

Variability of methods is linked to the issue of being quantitative or qualitative. However, the categories provided in Chap. 1 for methods did not match the definition of some contributing authors. They added methods that we did not include in Chap. 1, such as modelling workshops, literature, data bases, and random sampling. We believe this is mainly due to a lack of agreement on terminology and what a specific method precisely includes. Rather than get bogged down in semantics, in this book we have kept the diversity of methods as the authors of the individual chapters originally specified them.

The principle form of added value gained by describing cases with this framework is that it is made explicit that the model characterization and parameterization process needs to derive model assumptions cautiously from data and available information. Hence, transparency is a critical principle in this process. The framework forces the modeller to make the relations of an agent-based model to its empirical basis explicit and to be clear about the form of any (inevitable) interpretational filters that are present. Each contributing author reveals the relevant data sources and how they were implemented.

13.3 Robustness of Case Distinction

An essential cornerstone in the approach described in Chap. 1 is the acknowledgement that modellers can face very different situations. Due to the differences in conditions that characterise these situations we assume that generic recommendations cannot be developed. Instead, we aim for recommendations on basis of these distinct cases. The efficacy of these cases to distinguish situations needs to be tested.

Robustness of case distinction is strengthened if multiple examples of existing empirical agent-based models confirm a particular case. This Volume provides three examples for case 1 and two examples for case 7. This also paves the way for the opportunity to compare approaches for similar modelling situations.

The process of editing this book is a major step in developing a robust distinction of cases and testing them. The cases depicted in Chap. 1 changed substantially from an earlier version (Smajgl et al. 2011), which is a result of engaging with a larger set of empirical agent-based models and an attempt to classify them against the backdrop of model characterisation and parameterisation. For each of the described cases a particular sequence of steps was identified. While some of these variations in routing through the CAP framework seem minor, the differences are sufficient to point at different methodological recommendations.

Most authors that contributed to this Volume had no difficulty in mapping their example into the decision tree (Fig. 1.2 in Chap. 1). However, at times the sequence of criteria the decision tree employs was questioned. This might suggest potential for a more effective structure of the decision tree. Nevertheless, based on the modelling examples of this Volume the decision tree we developed proved to be largely effective.

The size of the population (number of agents) and the availability of data constitute key dimensions for the structure that distinguishes the cases we identified. We also discussed other aspects that characterise the modelling situation, in particular the modelling goal and the engagement process. While the final version of the decision tree is focused on data sources and their availability we believe that stakeholder related criteria could play a role in revising the decision tree. Whenever interactions with stakeholders are planned during the design process, the format needs to suit stakeholders' representations, so that these can understand and accept the outcomes of simulations. This can orientate methods that can be used. In Sect. 5 of this Chapter we revisit this argument.

The next two Sections discuss the scope of methods used in these chapters, with the following Section investigating the relation between case and methods.

13.4 Diversity of Methods

Table 13.1 below describes all the types of methods mentioned in Chaps. 2–12. Several insights emanate from this overview beyond the mere acknowledgement of the methodological diversity available to empirical agent-based modellers.

Expert knowledge is everywhere, even with the restricted definition of experts provided in Chap. 1. Experts are most often used in model characterization, but may be used in each step. To a lesser extent literature review has the same feature, which is not surprising since literature reviews are a specific format of expert knowledge.

Considering the relevance of expert knowledge it seems critical to understand potential fallback options for situations in which adequate expert knowledge is not available. Simulation through a random walk in the parameter space, linked to a sensitivity analysis is a possible quantitative technique to improve the robustness of parameter values, as shown by Gao and Hailu in Chap. 2 and Janssen in Chap. 10. Elements of uncertainty in the characterisation might also be transformed into a model parameter when several structures seem plausible and cannot be arbitrated on an empirical basis.

The sample of empirical models presented in this Volume suggests that some methods seem to dominate particular steps. Participant Observation is more often

Table 13.1 Methods used for each step in the CAP process

	M1	M2	M3	M4	4a	4b	M5
Expert knowledge	Gr Sa Ga LP IV Sm Qa	IV Qa	IV Sm Qa	Er IV	IV Sm Qa	IV Qa	IV Sm Qa
Literature review	Gr IV Hu Qa	Er IV Qa	IV Qa	Er	IV Qa	IV Qa	IV
Participant observation	Gr Ga IV Qa	Qa	Qa		Qa	IV Qa	
Theory	Er Sa Ja	Ja	Er Sa Qa				
Focus group	Gr						
Lab experiment	Ja		Ja				
Logbook	Ga						
Census	Sa Hu	IV Hu					IV Sm
Survey		Er Sa Ga LP IV Sm	LP IV				
Observations	Qa	Ja Qa	Qa		Qa	Qa	
Expert workshop		Er	Er				
Interviews	Ja IV	LP Qa	Gr Do Ga LP IV Sm Qa	Er			

Table 13.1 (continued)

	M1	M2	M3	M4	4a	4b	M5
Data bases	Gr	Gr	Gr	Sa			Sa
	Hu		Do				
			Hu				
Role playing game	IV	LP	LP				
			IV				
Modeling workshop			LP		Sm		
			Sm				
Field experiment			IV				
Time series			IV	Er			
Sensitivity analysis				Ja			
Clustering				IV	IV	IV	
					Sm		
Regression					IV	IV	
Correlation					IV	IV	
Dasymetric mapping					IV		
Mapping							Er
Random sampling							Ga
Cloning							IV
Monte Carlo							IV

Do: Dorscher et al., *Er:* Ernst, *Ga:* Gao & Hailu, *Gr:* Grozev et al., *Hu:* Huet et al., *IV:* Gray et al., *Ja:* Janssen, *LP:* Le Page et al., *Qa:* Geller, *Sa:* Sahrbacher et al., *Sm:* Smajgl & Bohensky

used for model characterization (M1), while surveys are typically used to elicit attribute data (M2) and interviews to elicit behavioural data (M3). Surveys and interviews are used in more than half of all cases, for M2 and M3 respectively. This can be explained by the greater adaptability of interviews compared to surveys. Behavioural data are more difficult to format *ex ante* and the quality of answers increases if exchange is possible between the interviewer and the interviewee. Asking people about their likely intentions means regularly facing responses that introduce new conditions for certain behaviours. Thus employing interviews allows for identifying conditions that have not arisen in previous design steps, which allows for implementing more realistic agent rules.

Concerning M1, participant observation is still less frequent than expert knowledge in characterising agents and other entities of the model. Participant observation is a rather time consuming technique, potentially difficult or impossible to implement during the lifetime of the model, and is bound to interpretational biases. However, it gives a hands-on perspective on the target system, which is useful to characterise a model or even to get information to elicit attribute and behavioural data. However, compared to later steps Participant Observation is mostly used for M1.

Surprisingly, census data and time series data come second to these qualitative methods, in total as well as for any specific step. Most empirical agent-based models included in this Volume elicit primary data and combine them with existing data. Only the case of Huet et al. in Chap. 8 uses solely census data for M2 and M3. This indicates that even cases with large human populations need to pay the effort of collecting primary data. Databases for large populations are either not available due to issues around human ethics and privacy, due to prohibitive development or maintenance costs, or due to format incompatibility.

Primary data can also be collected by experimental methods (field and lab experiments, role playing games), which create situations where a sample of agents is asked to perform a role within a set of constraints. They are well suited for collecting behavioural information but remain rather marginal as only a few examples employed these methods. Once used, these methods also provide information regarding human attributes and add to the characterisation of the model.

The final steps in the sequence (M4 through M5) are less explicit in most cases, and no method appears clearly dominant. Some cases feature the need to go back to existing data or existing knowledge in association with the use of statistical methods. Finally, some empirical agent-based modellers develop their own algorithms to generate virtual population fitting statistics found in official data bases.

In synthesis, all Chapters employ a combination of methods, as recommended by Poteete et al. (2010). This entails overcoming biases due to availability and accessibility of data which may influence the outcome of an empirical agent-based modelling process.

The application of multiple methods is clearly a positive observation. However, transparency is sometimes lacking. For instance, while all examples mention the use of expert knowledge, literature review or theory at some point, the reader is rarely provided with the rationale why a specific theory is chosen for a particular context. Selection of data in large databases, interpretation of observations or answers received in interviews are also often not fully explained. This introduces barriers for the reader to replicate the process as some of these implicit choices could be done differently, leading to different model assumptions.

Clearly, model development faces always constraints. There is no situation in which everything a model requires can be sourced from the empirical situation at a reasonable cost. Further, the modelling process is always implemented with a particular focus, which emerges from the modelling goal. This focus translates in a prioritisation of some data needs. Therefore, theory or literature is largely utilised to develop underpinning assumptions and to fill data gaps. In other words, theory provides a starting point to structure the system as well as to 'patch-up' gaps a modeller faces.

If we want to learn as a community from each others' experience we have to improve the transparency for how we use not only empirical data but also theory. Ultimately, model development always faces constraints. This emphasises the need to list the details for data-related and theory-related decisions and processes when implementing the CAP framework.

An additional transparency challenge arises from the use of pre-established models. The suitability of these previously designed models is not always discussed. For example Sahrbacher et al. refer to a "common assumption in agricultural economics", Gao and Hailu refer to principles of utility maximization. It is important that these implicit assumptions are made explicit and their suitability is discussed for the actual modelling context.

However, the framework-based description of empirical agent-based models allows repetition of a particular sequence of methods that employs similar empirical knowledge. Such an "eM2eM" experiment, that is still to be done, would provide the means to assess impacts on the model design outcome.

Regarding data, granularity or scale chosen to design the model will depend also on technical considerations or available data, constraining the representation of processes. As pointed out by Huet et al. in their Chapter, there are implicit models behind the format of databases. Such metadata also integrate expert knowledge. Depending on the mode of model use this issue can gain increasing relevance.

13.5 Modelling Goal and the Choice of Methods

The overarching goal of this book is to provide modellers with a set of methodological and process-related recommendations on CAP. However, the number of case studies presented here does not yet allow for inference around robust rules for method selection as a function of the type of empirical agent-based modelling situation. Newcomers to empirical agent-based modelling should rather see this Volume as a set of examples and experiences, which can help guide their methodological choices. This book allows the reader to identify the case that best describes their own modelling situation and to investigate whether the methods used in the matching example are suitable for their own case. The framework provides the main structure by defining steps M1 to M5. Empirical agent-based model designers should specify which steps are needed and identify the suitable methods for each step. In this section, we discuss the adequacy of methods in the context of modelling goals.

The use of models and simulation outcomes can have an important influence on the choice of methods. Whenever the engagement process includes the presentation of simulation results to end users, the engagement process should consider the system representation these end users hold and the related beliefs or habits of what type of representation is convenient. This need for legitimizing simulation outcomes through well accepted intermediary steps in the modelling process is increased when the simulation and modelling goal is influential and assumptions are controversial. This implies that the purpose of the modelling and the related modelling process can have implications for the choice of CAP methods. We distinguish four categories of modelling objectives:

- better understand components of a model,
- better understand some specific behavioural patterns,
- facilitate learning among decision makers and decision influencers, and
- provide decision support to policy makers.

Most examples provided in this Volume seem to follow the second objective, with a tendency towards the third. This diversity of objectives has two consequences for the designer of empirical agent-based models, in particular when employing expert knowledge. For the first two objectives, empirical agent-based modelling processes seem focused on specific model components or behaviours for which empirical data is elicited. Other parts of the model or behaviours with lower priorities are informed by commonly accepted assumptions or expert knowledge. For example, Gao and Hailu in Chap. 2 aim at specifying empirically behavioural patterns of trip timing decisions for recreational fishing. They consider that agents have a standard maximization of utility profile and gather data on fishing practices with surveys and use of logbooks.

When moving to the third and more so to the fourth modelling objective transparency and precision in data sources and data translation gain importance. Without allowing stakeholders to assess the suitability of the data the risk increases of stakeholders rejecting the model and model results due to black box characteristics. Hence, learning cannot be facilitated and decisions cannot be informed. The objective of facilitating learning through the exploration of potential scenarios requires, for instance, the use of random processes to generate noise and test the robustness of an institutional context, as it is done in the case of Le Page et al.

We assume that future testing will identify new cases, help to further expand on existing cases or to potentially suggest the merger of existing cases. Over time we hope to see a robust set of cases evolving, which will give newcomers effective guidance in identifying tested recommendations for a situation they are facing in developing an empirical agent-based model.

13.6 Good Luck with Your Models!

First of all this framework may serve as a suitable guideline when designing a modelling process and describing it for communication in the modelling community. The sample of cases is not yet sufficient to test the performance of methods in particular situations or to recommend specific sets of methods for specific cases. However, further sharing and testing of experiences described with this standardised framework might end up with recommendations for a given modelling situation.

13.7 Perspectives: Towards an Empirical Agent-Based Modelling Community?

The process of editing this book shows that there is not a "community" of empirical agent-based model designers. There are many divergences and the classic dichotomy between qualitative/quantitative, big and representative vs. small and ad hoc is still very present.

However, we believe this book shows that such a community could emerge. All authors share the same attention to derive their model design and parameterisation from empirical information, while understanding biases hidden in secondary data they are bound to use. This book should be understood as a way to reach out and connect to more modelling groups in order to improve the robustness of empirical agent-based modelling. We hope that the framework can be further improved and that the classification tree of empirical agent-based modelling cases can be further tested. If all cases have multiple examples described in the same structure the comparative discussion will allow derivation of design principles for robust empirical agent-based modelling and, ultimately, a wider acceptance of empirical agent-based modelling.

References

Poteete AR, Janssen MA et al (2010) Working togeher: collective action, the commons, and multiple methods in practice. Princeton University Press, Princeton

Smajgl A, Brown DG et al (2011) Empirical characterisation of agent behaviours in socio-ecological systems. Environ Modell Softw 26(7):837–844

Printed by Publishers' Graphics LLC
JCIMO130926.15.14.5